HEALTH PROTECTION FROM CHEMICALS IN THE WORKPLACE

ELLIS HORWOOD SERIES IN APPLIED SCIENCE AND INDUSTRIAL TECHNOLOGY

Series Editor: Dr D. H. SHARP, OBE, former General Secretary, Society of Chemical Industry; former General Secretary, Institution of Chemical Engineers; and former Technical Director, Confederation of British Industry

This collection of books is designed to meet the needs of technologists already working in the fields to be covered, and those new to the industries concerned. The series comprises valuable works of reference for scientists and engineers in many fields, with special usefulness to technologists and entrepreneurs in developing countries.

Students of chemical engineering, industrial and applied chemistry, and related fields, will also find these books of great use, with their emphasis on the practical technology as well as theory. The authors are highly qualified chemical engineers and industrial chemists with extensive experience, who write with the authority gained from their years in industry.

Published and in active publication

PRACTICAL USES OF DIAMOND
A. BAKON, Research Centre of Geological Technique, Warsaw, and A. SZYMANSKI, Institute of Electronic Materials Technology, Warsaw
NATURAL GLASSES
V. BOUSKA *et al.*, Czechoslovak Society for Mineralogy & Geology, Czechoslovakia
INTRODUCTION TO PERFUMERY: Technology and Marketing
TONY CURTIS, Senior Lecturer in Business Policy and International Business, Plymouth Business School, and DAVID G. WILLIAMS, Independent Consultant, Perfumery and Director of Studies, Perfumery Education Centre
POTTERY SCIENCE: Materials, Processes and Products
A. DINSDALE, lately Director of Research, British Ceramic Research Association
THE HOSPITAL LABORATORY: Strategy Equipment, Management and Economics
T.B. HALES, Arrowe Park Hospital, Upton, Wirral
MEASURING COLOUR: Second Edition
R. W. G. HUNT, Visiting Professor, The City University, London
LABORATORIES: Design, Safety and Project Management
Editor: T.J. KOMOLY, ICI Engineering, Winnington, Cheshire
PAINT AND SURFACE COATINGS: Theory and Practice
Editor: R. LAMBOURNE, Technical Manager, INDCOLLAG (Industrial Colloid Advisory Group), Department of Physical Chemistry, University of Bristol
HEALTH PROTECTION FROM CHEMICALS IN THE WORKPLACE
Editor: Dr P. LEWIS, Senior Executive, Occupational Health, Chemical Industries Association, London
FERTILIZER TECHNOLOGY
G.C. LOWRISON, Consultant, Bradford
NON-WOVEN BONDED FABRICS
Editor: J. LUNENSCHLOSS, Institute of Textile Technology of the Rhenish-Westphalian Technical University, Aachen, and W. ALBRECHT, Wuppertal
REPROCESSING OF TYRES AND RUBBER WASTES: Recycling from the Rubber Products Industry
V.M. MAKAROV, Head of General Chemical Engineering, Labour Protection, and Nature Conservation Department, Yaroslav Polytechnic Institute, Russia, and VALERIJ F. DROZDOVSKI, Head of the Rubber Reclaiming Laboratory, Research Institute of the Tyre Industry, Moscow, Russia
ACCIDENTAL EXPLOSIONS: Volume 1: Physical and Chemical Properties
LOUIS A. MEDARD, formerly Head of the Laboratory in the French Government Explosive Branch
ACCIDENTAL EXPLOSIONS: Volume 2: Types of Explosive Substances
LOUIS A. MEDARD, formerly Head of the Laboratory in the French Government Explosive Branch
PROFIT BY QUALITY: The Essentials of Industrial Survival
P.W. MOIR, Consultant, West Sussex
EFFICIENT BEYOND IMAGINING: CIM and its Applications for Today's Industry
P.W. MOIR, Consultant, West Sussex
COLOUR DYNAMICS AS A SCIENCE
ANTAL NEMCSICS, Professor at the Technical University of Budapest, Hungary
BY-PRODUCTS AND WASTE MATERIALS IN FAT TECHNOLOGY
HENRYK NIEWIADOMSKI, Technical University of Gdansk, Poland, and HANNA SZCZEPANSKA (deceased), formerly of the Institute of Industrial Chemistry, Warsaw, Poland
TRANSIENT SIMULATION METHODS FOR GAS NETWORKS
A.J. OSIADACZ, UMIST, Manchester
THE MECHANICS OF WOOL STRUCTURES
R. POSTLE, G.A. CARNABY and S. de JONG, Department of Textile Technology, University of New South Wales, Kensington, Australia
QUALITY ASSURANCE: The Route to Efficiency and Competitiveness, Third Edition
L. STEBBING, Quality Management Consultant

series continued at back of book

3 0116 00416 5005

This book is due for return not later than the last date stamped below, unless recalled sooner.

1 6 MAY 2001 LONG LOAN	2 0 MAY 2003 LONG LOAN
2 5 MAY 2001 LONG LOAN	
- 7 MAY 2002 LONG LOAN	

HEALTH PROTECTION FROM CHEMICALS IN THE WORKPLACE

Edited by:
Dr P. LEWIS,
Senior Executive, Occupational Health,
Chemical Industries Association, London

ELLIS HORWOOD
NEW YORK LONDON TORONTO SYDNEY TOKYO SINGAPORE

First published in 1993 by
ELLIS HORWOOD LIMITED
Market Cross House, Cooper Street,
Chichester, West Sussex, PO19 1EB, England

 A division of
Simon & Schuster International Group
A Paramount Communications Company

© Ellis Horwood Limited, 1993

All rights reserved. No part of this publication may be reproduced, stored in a retrieval system, or transmitted, in any form, or by any means, electronic, mechanical, photocopying, recording or otherwise, without the prior permission, in writing, of the publisher

Printed and bound in Great Britain
by Hartnolls, Bodmin

British Library Cataloguing in Publication Data

A catalogue record for this book is available from the British Library

ISBN 0–13–388240–3

Library of Congress Cataloging-in-Publication Data

Available from the publisher

Table of contents

Preface vii

1 **The UK regulatory scene** 1
  Dr P. Lewis, CIA

2 **The international regulatory scene** 19
  Dr J. R. Jackson, Monsanto Services International

3 **Non-governmental organizations** 39
  Dr J. R. Jackson, Monsanto Services International

4 **Occupational health services** 57
  Dr C. P. Juniper, Unilever

5 **Occupational health auditing** 71
  Dr M. K. B. Molyneux, Shell UK

6 **Hazard assessment** 95
  Dr F. M. B. Carpanini, BP International

7 **Assessment of health risks** 105
  Mr J. Holding and Mr N. Budworth, Bayer UK

8 **Ventilation systems** 127
  Mr J. G. Lyons, Unilever

9	**Personal protective equipment** Mr P. J. Turnbull, Associated Octel	137
10	**Occupational exposure limits** Mr A. M. Moses, ICI	157
11	**Sampling strategies** Dr I. G. Guest, Glaxo	173
12	**Acute poisoning** Dr I. J. Lambert, Shell UK	213
13	**Dermatitis** Dr P. C. Nicholson and Dr A. Chojnacki, Glaxochem	225
14	**Chronic health effects and chemical control** Dr P. A. Martin, Albright and Wilson	243

Appendix I: Glossary of acronyms — 269

Appendix II: Some health publications issued by CIA — 275

Index — 281

Preface

This book is one in a series of works produced by Ellis Horwood on applied science and industrial technology. It deals with a number of key topics of current interest concerning the protection of the health of people at work from the effects of hazardous chemicals. The topics are presented in a logical order and range from regulatory requirements, through those being addressed by major, non-governmental bodies, to purely voluntary guidance.

The book covers such areas as the provision of health services in industry, the auditing of occupational health, the evaluation of chemical hazards, the assessment of health risks and approaches for the prevention or minimization of exposure to chemicals. It also addresses the principal health effects which can arise from short- or long-term exposure to chemicals.

The chapters for the book have been written by occupational health specialists working for member companies of the Chemical Industries Association (CIA). All involved have considerable experience in producing health and safety guidance for the industry and have been key contributors on those topics for CIA. Whilst some material is necessarily fairly complex, most has been written at a level which can be readily appreciated by non-specialists.

The text is aimed at management and staff in the chemical industry and other industries where chemicals are used, particularly people on the shopfloor who handle potentially harmful chemicals. It is hoped that they will find something of value to their own working conditions, in the way of practical advice which can be readily applied, either directly or by way of approaches through health professionals, consultants, safety representatives, the Factory Inspectorate and others. The Editor also hopes that the book will find a valuable place both as background reading for training seminars and programmes and as an aid to discussion therein.

It is hoped that the reader will also find helpful the 'Glossary of acronyms' (Appendix I) which defines all important acronyms used in the book. Principal references to places in the text where these acronyms appear can be found from the index.

The book includes numerous references to various guidance booklets issued by CIA. Brief summaries of some of these are given in Appendix II together with an indication as to how they can be obtained. These cover key aspects of the COSHH Regulations as well as other important health issues. These booklets, in fact, represent only a small part of the considerable volume of codes of practice, guidance booklets and other publications issued by the association.

Policy on issues in health, safety and the environment for the CIA is determined through its Chemical Industry Safety, Health and Environment Council (CISHEC). This is composed of directors, senior managers and professional specialists representing the member companies of CIA. This body seeks to raise the standards of occupational health and safety, product safety, environment and distribution by disseminating information on good practice. CISHEC speaks, on behalf of the association's members, to numerous national and international bodies, such as the Government, the Health and Safety Commission and Executive, local authorities, learned societies, professional institutions, the media and the public.

A major preoccupation of the CIA, in recent years, has been the development of an important initiative, known as Responsible Care. This is an umbrella programme seeking to achieve continuous improvement in the performance of member companies in the fields of health, safety, environment, product safety, distribution, emergency response and relations with the public, and to enable companies to demonstrate that these improvements are, in fact, taking place. In this programme, chemical companies are sharing their experience in tackling problems in these fields. Responsible Care is concerned with performance; it is not a public relations exercise.

Although it is a voluntary programme, all members of CIA had committed themselves, by July 1992, to upholding the guiding principles of Responsible Care, after which date, it became a condition of membership of the association. Responsible Care is rapidly gaining the attention of member companies, not only of CIA but of other chemical industry federations throughout Europe, America, Canada, Australia and other parts of the world.

One important new Responsible Care publication on 'Safety, occupational health and environmental protection auditing', issued by CIA in 1991, outlined, for the first time, a generally agreed approach to health auditing. This laid the foundation for the production of a questionnaire, designed to investigate the level of attention paid to health protection within companies, which was recently launched throughout the CIA membership. It is hoped that this will point the way towards areas where improvements can be made, in future.

The launch of this book is also timely because of the European Community's recent campaign on such matters. The European Commission is seen as having a particularly important role to play in dealing with the serious problem of occupational accidents and illnesses. The human and economic cost of these is considerable. For example, of the 150 million workers in the Community, almost 10 million fall victim every year, 8000 of them fatally, to incidents, accidents or illnesses at work. Estimates put the annual compensation for occupational accidents and illnesses at 20 billion ECU, not counting indirect costs, which are substantial, but difficult to quantify.

To increase awareness of health and safety issues and to stress the economic as well as the social importance of high standards of health and safety at work, the EC Council of Ministers designated 1st March 1992 to 28th February 1993, as the European Year of Health, Safety and Hygiene Protection at Work. The main themes of this were clean air at work, safe working practices, well being at work and measures against noise and vibration. The main targets for the year were the high risk sectors of agriculture, construction, fishing and mining as well as small firms and young people in training.

A major initiative of the HSE for the European Year was the UK's first Workplace Health and Safety Week, held in November 1992. At an awards ceremony to mark the achievements of the week, Dr J. T. Carter, Head of HSE's Field Operations Division, said that there should be 'no room for complacency about occupational accidents and ill-health in this country'. A recent HSE survey, conducted on 80 000 workers in the UK, suggested that 'the most common, occupationally-related illnesses are musculo-skeletal disorders, such as bad backs, with around 500 000 cases, in total; deafness, tinnitus and other ear conditions and stress/depression, each with 100 000; respiratory disease (including asthma) with around 80 000 cases; and skin disease with around 50 000 cases'.

CIA lent its support to the year by organizing a number of special events which included a presentation (in London, in July 1992) of gold safety awards to its qualifying members, a seminar on Chemical Control Legislation (in Birmingham, in September 1992), a seminar on 'What the Public Wants to Know' (in London, November 1992) and a major conference on 'Responsible Partnership' (in Harrogate, February 1993).

CIA views this book as a further contribution to the year. Clearly, this is not the first book on health protection from chemicals in the workplace and it will not be the last. It will, however, serve to illuminate some of the key topics of interest and concern to the association in the health area during the period of the EC Year of Health and Safety.

July 1993

Phillip Lewis
Editor

to my dear wife, Cirla,
for her encouragement and patience
at all times,
and to my son, Mark,
for his valued support

Following his graduation in chemistry at Exeter University in 1962 and having obtained his Ph.D. in physical organic chemistry from Birkbeck College, London University, in 1965, **Dr Lewis** carried out a research fellowship at the University of Southern California until 1966. He then worked for six years as a research chemist for BP at their research centre in Sunbury-on-Thames, Middlesex, helping to improve process technology on vinyl chloride and styrene polymerization. This was followed by four years at the CBI, working on European science and technology policy and the removal of technical barriers to trade. In 1976, Dr Lewis joined the Chemical Industries Association, where he has remained to the present, developing policy in occupational medicine, hygiene and toxicology. Now Senior Executive for Occupational Health, Dr Lewis was intimately involved with the formulation and advocacy of CIA's views during the development of legislation on the Control of Substances Hazardous to Health (COSHH) since its conception at the beginning of the 1980s.

1

The UK regulatory scene

Dr P. Lewis

SUMMARY

This chapter reviews the key elements of UK law on the control of workplace exposure to chemicals which can be harmful to health. The broad objectives for this are laid down in the Health and Safety at Work etc. Act 1974 and developed in more detail in the Control of Substances Hazardous to Health Regulations 1988 (COSHH).

The text outlines the main principles of COSHH as stated in its Regulations and amplified in the associated general Approved Code of Practice. Other related codes cover more specific areas such as carcinogens, vinyl chloride, fumigation, pesticides, etc. These requirements are backed up by HSE guidance documents EH 40 (on 'Occupational Exposure Limits'), EH 42 (on 'Monitoring Strategies for Toxic Substances') and others.

The chapter then describes the numerous guidance publications issued by the Chemical Industries Association (CIA) to assist its member companies (and others) handling hazardous chemicals on sound practices for compliance with the COSHH Regulations and for improving performance in health protection in accordance with the principles of Responsible Care.

Also touched on briefly are other regulations made under the Act on the control of lead and asbestos at work.

The chapter concludes with an outline of other legislation having an important bearing on this subject, in particular the Classification, Packaging and Labelling Regulations (CPL), the Reporting of Injuries, Diseases and Dangerous Occurrences Regulations (RIDDOR) and the Notification of New Substances Regulations (NONS).

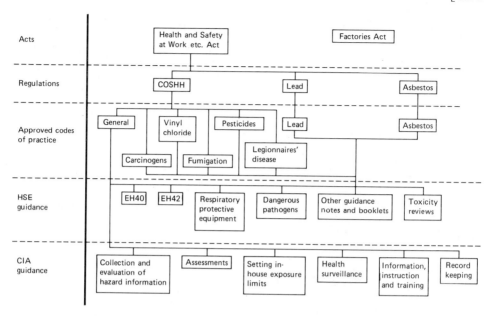

Fig. 1. UK legislation on the control of chemicals at work. Note: This figure is a plan of the subject matter covered by sections 1–4 of this chapter. It is purely illustrative of the relationship between the various kinds of documents referred to in the text; it is by no means complete.

FOREWORD

This chapter reviews the regulatory scene in the UK for protection from hazards to health from exposure to chemicals in the workplace and highlights the key elements which affect the mainstream chemical industry. The principal element in this is the COSHH legislation which forms the central theme of the review. It is not intended to be a comprehensive review nor should it be viewed in any way as complete. Also, the chapter does not cover legislation aimed primarily at the safety hazards of chemicals such as explosivity, flammability, etc. which are dealt with under major hazards legislation, etc.

Throughout this chapter, the reader should be aware that the UK regulatory scene is under considerable review and development. Much of the pressure for this change is now coming from the European Community's Social Action Programme. This contains nearly 50 measures designed to give effect to the declaration of intent contained in the Social Charter adopted in 1989. The programme has produced a flood of new directives which are being implemented during the current period. The EC regulatory scene is, however, covered in more detail in Chapter 2.

1. INTRODUCTION

Legislation on health and safety at work started to develop in the UK around the beginning of the nineteenth century. Important milestones include the Mines and

Quarries Act 1954, The Offices, Shops and Railway Premises Act 1963 and various items of legislation on factories which were finally consolidated in The Factories Act 1961. A revised guide on this last item was published in 1991 [1]. Note that a plan of the subject matter covered by sections 1–4 of this chapter is given in Fig. 1.

2. THE HEALTH AND SAFETY AT WORK ACT

The Health and Safety at Work etc. Act 1974 (the HSW Act) arose as a consequence of the inquiry into safety and health at work conducted by Lord Robens [2], and came into force in 1975. It is an enabling act outlining broad principles which are developed in detail in individual regulations and ap^{o}proved codes of practice on particular health and safety problems.

A comprehensive guide to the Act has been produced by the Health and Safety Executive (HSE) [3].

The Health and Safety Commission (HSC) was established in 1974 by the HSW Act as a tripartite body representing employers, trade unions and local authorities and independents.

The HSE, also created by the Act, is the executive arm of the Commission and includes a number of inspectorates such as Her Majesty's Factory Inspectorate (HMFI), the Agricultural, Mines, Quarries and Explosives Inspectorates and the Employment Medical Advisory Service (EMAS).

Responsibility for occupational health is vested in EMAS, which gives advice on all aspects of the subject to industry, general medical practitioners and various Government departments. An explanatory leaflet, describing the services and advice provided by EMAS is available from HSE [4].

3. THE CONTROL OF SUBSTANCES HAZARDOUS TO HEALTH (COSHH) LEGISLATION

COSHH is the most comprehensive piece of legislation on the control of chemicals at work adopted in the UK since the HSW Act. It was prompted largely by the need to implement the 1980 Framework Directive on the Control of Chemical, Physical and Biological Agents at Work (80/1107/EEC). Whilst the general duties for health and safety are outlined in the HSW Act, there was a need to spell out more clearly a rational approach to the control of risks from potential exposure to chemicals in the workplace. Section 2(1) of the HSW Act states that 'It shall be the duty of every employer to ensure, so far as is reasonably practicable, the health, safety and welfare at work of all his employees'.

The drafting of COSHH was achieved over a period of many years' consultation with industry, trade unions and others. CIA played a full part in discussions during this process and, through active participation of occupational health professionals from its member companies, helped to develop both the principles and the detailed wording of key parts of the legislation.

3.1 The COSHH Regulations
Much of the approach adopted in the COSHH Regulations 1988 [5] was modelled on the Control of Lead at Work Regulations 1980 [6] . Lead and asbestos are excluded from COSHH as they are the subject of their own, separate regulations. (See section 4 below).

The introduction of the COSHH Regulations was accompanied by the repeal of numerous sets of regulations on specific chemicals or groups of chemicals which had been made years before. A full list of these is given in Schedule 9 of the Regulations but notable among them were:

— the Chemical Works Regulations 1922,
— the Chromium Plating Regulations 1931, and
— the Carcinogens Substances Regulations 1967.

Thus, COSHH sought to bring into a single instrument a legislative approach to the control of all substances potentially harmful to health. The regulations came fully into force on 1st January 1990 after a transitional period of 3 months to allow for the carrying out of risk assessments.

The approach embodied in COSHH (and its supporting codes of practice) is a logical framework of principles of good occupational hygiene practice and health care which allows companies the flexibility to decide on the appropriate action needed. In essence, the approach consists of:

(a) collecting and evaluating hazard information on the chemicals likely to occur in the workplace;
(b) assessing the risks to health in the various work activities performed by employees;
(c) deciding on appropriate measures to be applied to control or prevent those risks;
(d) ensuring proper maintenance of the control measures;
(e) monitoring the workplace atmosphere, if necessary, to ensure that contaminants are kept below harmful levels;
(f) carrying out appropriate health surveillance;
(g) providing appropriate information, instruction and training of workers; and
(h) keeping careful records of various of the above activities.

It should be noted that the assessment (under Regulation 6) requires (b) to be carried out as well as the steps (c)–(h) needed to be taken.

The broad requirements for the above strategy are laid down in the regulations themselves. Failure to comply with any of these regulations is, of itself, an offence under the law and can render a company liable to prosecution.

The COSHH Regulations are subject to a process of continuing review and amendment, much as a consequence of changes in European Community legislation. A review paper on the Impact of EC Legislation on COSHH has been published by the author of this chapter [7].

3.2 The COSHH General Approved Code of Practice
Appropriate means by which the objectives of the COSHH Regulations can be met are given in the HSC general Approved Code of Practice (ACoP) [8]. This describes

appropriate practical measures which can be taken based on good occupational hygiene practice.

As with all HSC approved codes, this ACoP has semi-legal status in that compliance with it is generally taken to mean that the regulations themselves are being observed. Failure to comply with any part of the code is not, of itself, an offence provided the requirements of the regulations are achieved by some other means. However, if a company is taken to court through a prosecution, etc., the onus will be on that company to satisfy the court that its alternative action was equally effective in meeting the regulations.

It should be noted that the regulations and the general ACoP are continually being amended to incorporate changes to the list of Maximum Exposure Limits (MELs) appearing in Schedule 1 and to reflect modifications required by EC legislation, etc. It is important, therefore, to ensure that one is dealing with current versions of these documents. To date, all these amendments have been introduced annually, with the amending regulations and ACoPs coming into effect on 1st January. Thus, for example, the COSHH (Amendment) regulations 1992 and the new ACoP (Fourth Edition) came into effect on 1.1.93

3.3 Other COSHH ACoPs

3.3.1 Carcinogens
This ACoP is a companion to the general one and must be read in conjunction with it. Indeed, it is published as part of the same document. Essentially, it emphasizes the requirements in the general ACoP which are particularly important when dealing with carcinogens. Particular emphasis is given here to the need to consider substitution of a carcinogen by a substance of lower risk to health.

The code applies to all substances† [and preparations] which have been classified as requiring to be labelled with the risk phrase, R 45, 'May cause cancer' [or R 49, 'May cause cancer by inhalation']. Of course, this may be only one of a number of risk and safety phrases which appear on the label of a particular product owing to the need to warn the user of other hazardous properties such as explosivity, flammability, corrosivity, etc.

Also covered by the Carcinogens ACoP are a number of specified work processes which are known to present potential carcinogenic risks but where the particular agents responsible are not known with certainty. The ACoP became of greater significance on 1.1.93 as it then covered the new Schedule 10 to COSHH which implemented the Carcinogens Directive [9].

3.3.2 Vinyl chloride
This ACoP [10] came into force at the same time as the regulations and general ACoP and relates solely to occupational exposure to vinyl chloride monomer. It amplifies the principles of the general ACoP specifically for this substance. A group

---

† Note. Items within square brackets do not appear in the Third Edition of the Carcinogens ACoP but are expected to appear in the Fourth Edition.

of health professionals from member companies of CIA was closely involved in the drafting of this ACoP.

Repeated exposure in excess of the exposure limits for vinyl chloride can produce angiosarcoma of the liver and acro-osteolysis, a deterioration of certain bones, particularly in the fingers. However, the ACoP has somewhat limited application as there are few companies handling this substance.

3.3.3 Fumigation
The Fumigation ACoP [11] was also issued at the same time as the above ACoPs and describes procedures for controlling exposure to fumigation operations involving ethylene oxide, methyl bromide, phosphine and hydrogen cyanide. HSE guidance notes have also been issued on such workplace operations.

3.3.4 Other codes
Of the other codes which have been issued in relation to COSHH, that on Legionnaires' Disease is worthy of particular mention. Introduced in 1992 [12], this sets out a framework for preventing outbreaks of the disease in relation to requirements under both the HSW Act and the COSHH Regulations. It requires employers to identify and assess sources of risk from Legionella and draw up a scheme for preventing or minimizing the risk. Guidance and a video for use in training are also available.

An ACoP on non-agricultural pesticides, issued in 1991 [13], provides advice to users of these substances under the COSHH Regulations. It is aimed at people using such materials in amenity horticulture, wood preservation, the application of anti-fouling paints, etc., and serves as a complement to the code of practice on the safe use of pesticides on farms, etc. issued the previous year. The latter ACoP is interesting as it is a joint code covering both the Control of Pesticides Regulations 1986 and COSHH.

3.4 HSE guidance relating to COSHH
A detailed treatment of the COSHH Regulations is outside the scope of this chapter and is therefore not given here. An enormous volume of interpretive guidance has been produced since the coming into force of this legislation, a full list of which can be obtained from the HSE.

A particularly valuable guide to the whole of the regulations is that produced originally for local authority inspectors and subsequently made available to the public as an open learning course [14].

HSE has produced a guide to assessments [15] and health surveillance under COSHH [16] which were of great help in overcoming industry's difficulties in setting about these important tasks for implementing the regulations. HSE has also issued numerous other guidance booklets, guidance notes and leaflets, key items of which are mentioned below. These have been extremely helpful in clarifying the understanding on important issues.

Two guidance documents, in particular, deserve mention in the context of COSHH, namely Guidance Booklet EH 40: 'Occupational Exposure Limits' [17] and Guidance Note EH 42: 'Monitoring Strategies for Toxic Substances' [18].

3.4.1 EH 40
This contains a very useful introductory text describing the procedure used by the Commission's Advisory Committee on Toxic Substances (ACTS) and its Working Group for the Assessment of Toxic Chemicals (WATCH) for the setting of occupational exposure limits. Also given is useful guidance on the application of the limits for exposure both to single substances and to mixtures.

Tables list current values for all adopted Maximum Exposure Limits (MELs) and Occupational Exposure Standards (OESs). Also given are proposals for new OESs to come into effect the following year, giving interested parties the opportunity to comment on them. Various appendices give useful information on the calculation of exposure levels for specified reference periods, notes on limits for particular groups of chemicals, a list of all substances requiring the R 45 risk phrase and references to measurement methods published by the HSE. The guidance booklet is revised annually and appears at the beginning of the year together with the new ACoP and amending COSHH Regulations.

3.4.2 EH 42
This HSE guidance note provides practical advice on the monitoring of workplace atmospheres to ensure that contaminants are kept below harmful levels.

It discusses the factors which can affect the concentrations of hazardous substances in workplace atmospheres and outlines a structured approach to evaluating exposure through an initial appraisal, a basic survey, a detailed survey through to routine monitoring.

3.4.3 Other HSE guidance
Amongst other HSE guidance worthy of mention here is an excellent guide on the use of respiratory protective equipment (RPE) used for controlling workers' exposure to airborne substances likely to damage their health [19]. This covers breathing apparatus, filtering respirators and air-fed equipment such as hoods, helmets, visors and blouses.

The document sets out criteria for suitability and categories of RPE available for use. Under regulations on COSHH, lead and asbestos, RPE should be used only where other methods of controlling exposure are not reasonably practicable.

A particularly important item of HSE guidance here is that on dangerous pathogens, (i.e. bacteria, fungi, viruses and parasites) a revised document on which was issued in 1990 produced by the Commission's Advisory Committee on Dangerous Pathogens [20]. First published in 1984, the guidance has become a standard reference document for people working with micro-organisms in laboratories. It classifies pathogens into four hazard groups.

HSE has also produced a large number of guidance notes under the headings medical, environmental hygiene, chemical safety, plant and machinery and general. Although too numerous to list completely here, the following titles are worthy of particular mention:

- 'Pre-employment Health Screening' (MS 20),
- 'Health Aspects of Job Placement and Rehabilitation' (MS 23),
- 'Health Surveillance of Occupational Skin Disease' (MS 24),

as well as guidance on occupational asthma (MS 25), dusts (EH 44), biological monitoring (EH 56) and a large number of items on particular hazardous chemicals or groups of chemicals.

HSE has also published a considerable number of reviews on the toxicity of chemicals considered by ACTS for the setting of occupational exposure limits and other control procedures.

3.5 Guidance produced by the Department of Health
The Department of Health has issued a revised edition of their Guidelines for the Evaluation of Chemicals for Carcinogenicity [21]. It addresses the evaluation of data from epidemiology, structural chemistry, metabolic studies, short-term mutagenicity tests and long-term animal testing.

This is a companion booklet to others produced on toxicity and mutagenicity.

3.6 COSHH guidance produced by trade associations
A number of trade associations have produced guidance—with encouragement and assistance from HSE—on important aspects of regulations affecting their industries.

When COSHH came into force, many member companies of CIA had already initiated major programmes to implement the requirements despite, the additional resources needed both in terms of professional expertise and costs.

CIA believes that the legislation should be applied to great benefit in the chemical industry as well as the numerous industries which use substances hazardous to health.

In particular, CIA has produced a number of guidance booklets for its member companies, their customer companies and the various industry sectors which buy in and use chemicals. Topics covered by these booklets include the following.

3.6.1 Collection and evaluation of hazard information
This booklet provides guidance on the identification of substances to be controlled under COSHH. It expands on the criteria given in Regulation 2 for defining a substance hazardous to health and describes the sources of information useful for assessing the dangerous properties of substances.

3.6.2 Assessments
The fundamental requirement of COSHH is to carry out assessments of the health risks of employees from hazardous chemicals present at their places of work (Regulation 6). This CIA booklet offers guidance on a systematic approach to assessment for process units, workshops, laboratories, offices and stores.

3.6.3 Setting in-house exposure limits
This covers the need to set limits for substances that have not been assigned Maximum Exposure Limits (MELs) or Occupational Exposure Standards (OESs). A systematic

approach is described involving the review of health effects from information that can be obtained from various sources.

3.6.4 Health surveillance
COSHH Regulation 11 outlines the general principles of health surveillance and provides a framework for decision-making. This CIA booklet is designed to help employers decide when health surveillance is required, what kind of surveillance is needed, who should carry it out and what records should be kept.

3.6.5. Information, instruction and training
This booklet covers the requirement under Regulation 12 to provide employees with 'suitable and sufficient' information, instruction and training on their health risks and the precautions that should be taken. This includes the results of exposure monitoring and the collective results of health surveillance.

3.6.6 Record keeping
This summarizes the record-keeping requirements of the regulations and suggests the recording of other information that may be useful to demonstrate compliance.

3.6.7 Laboratories
The implementation of COSHH in laboratories is an important but difficult area because of the use of large numbers of chemicals for which little or no toxic hazard data may be available. The Royal Society of Chemistry, with the active involvement of CIA, has produced a valuable guidance booklet on this subject [22].

Booklets on other important areas of the regulations were in preparation at the time of writing. Details are obtainable from CIA [23].

Other CIA guidance publications relevant to COSHH that were produced prior to the introduction of the COSHH Regulations are to be revised. Notable among these are booklets on Carcinogens in the Workplace, Chemical Protective Clothing and Occupational Exposure Limits for Mixtures.

These booklets are only a few of the many publications produced by CIA not only to help industry meet its obligations under legislation but also to extend good practice to a wider section of industry. Further information on the booklets is given in Appendix II at the end of the book.

Recently, considerable enthusiasm has arisen within CIA for the Responsible Care programme which aims to promote improved standards of performance in health, safety and the environment. Fundamental to this is a signatory commitment by chief executive officers of all CIA member companies to a set of guiding principles (see Fig. 2). Companies are now expected to comply with such guidance booklets and other codes of practice produced by the Association. In order to raise standards of health protection, it is important to improve attitudes to, and awareness of, good practice by individuals at all levels in companies. This cannot be achieved solely by the imposition of excessively detailed legislation.

 Responsible Care

HEALTH SAFETY AND ENVIRONMENT
GUIDING PRINCIPLES

As a member of the Chemical Industries Association this company is committed to managing its activities so that they present an acceptably high level of protection for the health and safety of employees, customers, the public and the environment.

The following Guiding Principles form the basis of this commitment :-

- Companies should ensure that their health, safety and environment policy reflects the commitment and is clearly seen to be an integral part of their overall business policy.

- Companies should ensure that management, employees at all levels and those in contractual relationships with the Company are aware of their commitment and are involved in the achievement of their policy objectives.

- All Company activities and operations must be conducted in accordance with relevant statutory obligations. In addition, Companies should operate to the best practices of the industry and in accordance with Government and Association guidance.

In particular, Companies should :-

- Assess the actual and potential impact of their activities and products on the health and safety of employees, customers, the public and environment.

- Where appropriate, work closely with public and statutory bodies in the development and implementation of measures designed to achieve an acceptably high level of health, safety and environmental protection.

- Make available to employees, customers, the public and statutory bodies, relevant information about activities that affect health, safety and the environment.

Members of the Association recognise that these Principles and activities should continue to be kept under regular review.

I, Chief Executive of ....

commit my Company to compliance with the above Guiding Principles.

....

Date

Fig. 2. Guiding principles for CIA's Responsible Care programme (see page 9).

CIA therefore commends the COSHH philosophy for application in downstream industries using hazardous substances in the UK, within the European Community and other parts of the industrialized world, particularly the USA and Japan.

4. REGULATIONS ON THE CONTROL OF EXPOSURE TO OTHER CHEMICALS AT WORK

4.1 Lead
The Control of Lead at Work Regulations 1980 [6] covers the need to control exposure to any form of lead or lead compounds such as dust and fumes in e.g. lead smelting, welding, etc., as well as work in the production of lead alkyls.

The regulations are backed up by an important ACoP [24] that gives detailed advice on precautions for the safe handling of these compounds. This has been amended to take account of changes required by the EC Directive on Lead [25]. For example, a tighter standard was introduced for the withdrawal of employees from work when their lead-in-blood levels exceed 70 mg/100 ml (compared with 80 mg/100 ml previously).

4.2 Asbestos
This class of compounds is also covered by its own set of regulations [26], separate from COSHH, which implemented the EC asbestos directive [27].

The regulations apply to any work involving asbestos or asbestos-containing products which may have the potential to cause exposure to asbestos dust. They are backed up by two ACoPs and various items of guidance.

5. OTHER IMPORTANT REGULATIONS

This section reviews other sets of regulations that have an important bearing on legislation on the control of chemicals in the workplace covered by sections 1–4 above. A plan of the subject matter covered by section 5 is given in Fig. 3.

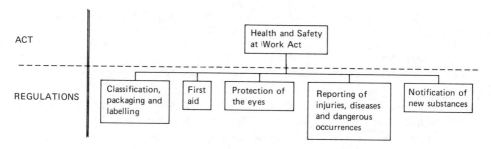

Fig. 3. Other UK legislation relevant to the control of chemicals at work. Note: This is a plan of the subject matter covered by section 5 of this chapter.

5.1 Classification, Packaging and Labelling Regulations

The Classification, Packaging and Labelling (CPL) of Dangerous Substances Regulations 1984 [28] implemented numerous requirements of the Sixth Amendment of the 1967 EC Directive on the Classification, Packaging and Labelling of Dangerous Substances [29]. These Regulations indicate the hazard categories and corresponding hazard warning symbols (See Fig. 4). The Regulations have since been amended several times to reflect various adaptations to the 1967 Directive.

Specific requirements for supply labelling are given in the Authorised and Approved List [30]. This indicates the substances which have internationally agreed labels. Substances not on this list must be classified and labelled by the supplier using the criteria in the Approved Code of Practice (COP 22) on Classification and Labelling of Substances Dangerous for Supply [31].

The CPL Regulations also require labels for dangerous substances carried by road. This part of the regulations is based on United Nations recommendations.

These regulations have an important bearing on COSHH since the latter has adopted the same rationale for defining a 'Substance Hazardous to Health' as 'very toxic', 'toxic', 'harmful', 'corrosive' or 'irritant', etc.

The European Commission's 'classification' programme also determines those substances which are required to be labelled 'May cause cancer' and thus automatically fall within the scope of the COSHH Carcinogens ACoP.

The CPL Regulations are to be extended to dangerous preparations to implement the EC directive on such preparations [32]. In effect, the Regulations will be withdrawn during 1993 and replaced by new regulations to implement the preparations directive.

5.2 First aid

Under the Health and Safety (First Aid) Regulations 1981 [33], made under the HSW Act, employers have to provide first aid, the nature of which depends on the types of hazards present, the medical services available, the number of employees and whether there is shift working.

An approved code of practice [34] and guidance notes are available which are based on several years' experience with the regulations. They reflect a change in emphasis towards assessment of risks of injury in each workplace to provide resources to match the needs. Thus, first aiders should be given training for the hazards they are likely to encounter in their workplaces.

Both the regulations and ACoP were under review at the time of writing.

5.3 Eyes regulations

The Protection of Eyes Regulations 1974 apply to factories and certain other premises covered by the Factories Act 1961 having any process likely to cause risks of eye injuries. Employers are required to provide such employees with suitable eye protectors such as goggles, visors, spectacles or face screens.

Certificates of approval and British Standards have been issued relating to specifications for eye protective equipment. A guidance booklet on this subject is available from CIA [23].

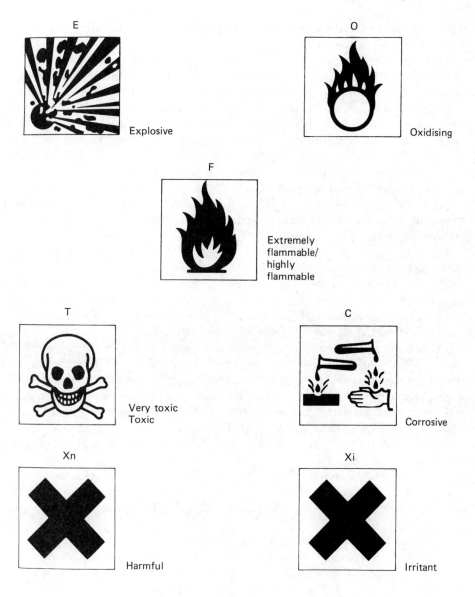

Fig. 4. Hazard warning symbols to be used under the Classification, Packaging and Labelling Regulations (see page 12). Symbols are black on an orange background.

5.4 Reporting of Injuries, Diseases and Dangerous Occurrences Regulations

These regulations (RIDDOR) [35] came into effect in 1986 to replace the Notification of Accidents and Dangerous Occurrences Regulations 1980 (NADOR). The effectiveness of the latter had been impaired by changes to the industrial injuries scheme which were introduced by the Social Security and Housing Benefits Act 1982.

RIDDOR also replaced the provisions for the notification of industrial diseases laid down in the Factories Act 1961.

RIDDOR is designed to generate reports to provide the enforcing authorities with information to help them in their accident and ill-health prevention activities by indicating where and how problems arise and by showing up trends.

Reportable diseases and related specific work activities are listed in Schedule 2 of the regulations. They are grouped under five headings: poisonings, skin diseases, lung diseases, infections and other conditions associated with exposure to particular chemical or physical agents.

An official guide to RIDDOR has also been produced [36].

After an initial period of operation—and at the time of writing—RIDDOR was under review particularly for the ill-health reporting provisions, owing to the need to consider inclusion of some items from the European Schedule of Industrial Diseases and other matters.

Prescribed industrial diseases
The National Insurance (Industrial Injuries) Act 1946 introduced the industrial injuries scheme under which the Government pays no-fault compensation for certain injuries and diseases. The act also established the Industrial Injuries Advisory Council (ITAC), an independent, statutory body of experts responsible, among other things, for making recommendations on changes to the related Schedule of Prescribed Industrial Diseases. The council reports to the Secretary of State for Social Security and, from time to time, recommends changes to the schedule for his consideration, in the light of particular investigations carried out. This schedule is thus covered by quite different legislation from RIDDOR and is a matter for the Department of Social Security, whereas RIDDOR is managed by the HSE.

Some 35 reports have been produced, to date, by the council, all but three being on prescribed diseases.

Notable among the reports, in recent years, include results of investigations into chronic bronchitis and emphysema, cancer of the larynx, occupational asthma, occupational lung cancer, halogenated aliphatic hydrocarbons and n-hexane.

5.5 Notification of New Substances Regulations
The notification scheme was established in the UK by the Notification of New Substances Regulations 1982 (NONS) as a result of the EC's Sixth Amendment Directive of the 1967 CPL Dangerous Substances Directive [37].

Substances must be notified if they do not appear in the European Inventory of Existing Commercial Chemical Substances (EINECS). This inventory lists about 100 000 substances which were on the market in the European Community at some time between 1.1.71 and 18.9.81.

The notification must include a dossier of information on the physico-chemical, toxicological and ecotoxicological properties of the substance. The dossier must be submitted to HSE before the product is placed on to the market so that its potential effects on human health and the environment can be assessed. The test methods required are generally those which have been developed under the auspices of the

Organisation for Economic Cooperation and Development (OECD). The methods have been adopted for the European Community in two directives [38, 39] which have, in turn, been adopted for the UK as HSC approved codes of practice for use in accordance with the notification regulations. The first set of tests, known as the 'base set', is required if the new substance is to be supplied in a quantity greater than 1 tonne per year, while the latter set includes additional tests for toxicity and ecotoxicity that may be needed when the quantity supplied exceeds 10 tonnes per year or a cumulative total of 50 tonnes.

For quantities below 1 tonne, a limited announcement is required. This includes information on the identity of the substance, the quantity to be supplied and the labelling requirements (from test data).

There are exemptions for certain substances, such as pesticides, which are covered by other regulations.

A useful, updated guide to these regulations has also been produced by the HSE [40].

All substances notified throughout the European Community between 18.9.81 and 30.6.90 have been published in the European List of Notified (New) Chemical Substances (ELINCS) [41].

6. PROPOSED NEW LEGISLATION

At the time of writing, proposals for a number of new regulations were being put forward to implement further new EC directives under the Social Charter. Important among these, in the context of this chapter, are the wide-ranging Management of Health and Safety at Work Regulations and Approved Code of Practice (aimed at implementing the 1989 Framework Directive on Measures to Encourage Improvements in the Safety and Health of Workers at Work) and the Personal Protective Equipment at Work Regulations and guidance (to implement the EC directive on the Use of PPE). These were introduced at the beginning of 1993 for the Single European Market.

7. CONCLUSIONS

This chapter has surveyed the main elements of UK legislation covering the health aspects of the control of exposure to chemicals at work. The chemical industry believes that broad objectives need to be laid down in Government legislation but that detailed means and procedures for achieving those objectives should be written in flexible terms either in guidance produced officially by the Government (in this case, the HSE) or for specific industry sectors, by industry itself through its trade associations and member companies.

The regulatory scene is continually changing and evolving, much being driven by the European Community. However, CIA directly and, through participation of its occupational health professionals, continues to play a full part in helping to shape new legislative requirements to allow the ends to be achieved by the most effective means in terms of resources and costs.

Aside from this, the Association is developing a major commitment to Responsible Care in the raising of standards of performance in health, safety and the environment backed up by an increasing volume of voluntary codes of practice and guidance material. Much of this energy derives from a voluntary, proactive sharing of industry experience and collaboration on good practice; this is believed to be more effective in the long term than detailed prescriptive diets of Government regulations.

ACKNOWLEDGEMENT

The author of this chapter wishes to record his appreciation of the kind assistance given by officials in the UK Health and Safety Executive and Government Department of Health in answering a number of queries on UK legislation.

REFERENCES

[1] 'Guide to the Factories Act 1961' (ISBN 0 11 885672 3), 1991, HMSO.
[2] 'Safety and Health at Work', Report of the Committee 1970–72 chaired by Lord Robens (SBN 10 1503407), Cmnd 5034, 1972, HMSO.
[3] 'A Guide to the HSW Act', Health and Safety Series Booklet, HS(R)6 (ISBN 0 11 883264 6), 1980, HMSO.
[4] 'An Introduction to the Employment Medical Advisory Service', HSE 5 (Revised), 1990, HSE.
[5] 'Control of Substances Hazardous to Health Regulations 1988', SI 1988 No. 1657 (ISBN 0 11 0876571), 1988, HMSO and subsequent amendments.
[6] 'Control of Lead at Work Regulations 1980', SI 1980 No. 1248 (ISBN 0 11 0072480), 1980, HMSO.
[7] 'The Impact of EC Legislation on COSHH', P Lewis, *Chemistry in Britain*, January 1991, pp 37–39.
[8] 'Control of Substances Hazardous to Health: General and Carcinogens Codes of Practice', Third Edition, L 5, (ISBN 0 11 885698 7), 1991, HMSO.
[9] 'Protection of Workers from the Risks related to Exposure to Carcinogens at Work', Directive 90/394/EEC, *Official Journal of the European Communities*, L 196, 26.7.90, pp 1–7, HMSO.
[10] 'Control of Vinyl Chloride at Work', Approved Code of Practice, COP 31 (ISBN 0 7176 0317 2), 1988, HMSO.
[11] 'Control of Substances Hazardous to Health in Fumigation Operations', Approved Code of Practice, COP 30 (ISBN 0 11 885469 0), 1988, HMSO.
[12] 'Prevention or Control of Legionellosis (including Legionnaires' Disease), Approved Code of Practice, L 8 (ISBN 0 11 885659 6), 1991, HMSO.
[13] 'Safe Use of Pesticides for Non-Agricultural Purposes', Approved Code of Practice, L 9 (ISBN 0 11 885673 1), 1991, HMSO.
[14] 'COSHH: An Open Learning Course', (ISBN 0 11 8854348), 1990, HMSO.
[15] 'COSHH Assessments: A Step-by-Step Guide to Assessments and the Skills Needed for It' (ISBN 0 11 885470 4), 1988, HMSO.
[16] 'Health Surveillance under COSHH', HSE Guidance Booklet (ISBN 0 11 885447 X), 1990, HMSO.

References

[17] 'EH 40/92: Occupational Exposure Limits 1992', HSE Guidance Booklet (ISBN 0 11 885696 0), 1992, HMSO, (Revised annually).

[18] 'Guidance Note EH 42: Monitoring Strategies for Toxic Substances' (ISBN 0 11 885412 7), 1989, HMSO.

[19] 'Respiratory Protective Equipment: A Practical Guide for Users', HSE Guidance Booklet HS(G)53 (ISBN 0 11 885522 0), 1990, HMSO.

[20] 'Categorisation of Pathogens according to Hazard and Categories of Containment', Second Edition, (ISBN 0 11 885564 6), 1990, HMSO.

[21] 'Guidelines for the Evaluation of Chemicals for Carcinogenicity' (ISBN 0 11 321453 7), 1991, HMSO.

[22] 'COSHH in Laboratories' (ISBN 0 85186 3191), 1989, Royal Society of Chemistry (to be revised).

[23] CIA Publications, Chemical Industries Association, Kings Buildings, Smith Square, London, SW1P 3JJ (Tel: 071-834 3399 Fax: 071-834 4469).

[24] 'Control of Lead at Work: Approved Code of Practice', (ISBN 0 11 883780 X), 1985, HMSO.

[25] 'Protection of Workers from the Risks related to Exposure to Metallic Lead and its Ionic Compounds at Work', Directive 82/605/EEC, *Official Journal of the European Communities*, L 247, 23.8.82, pp 12–21, HMSO.

[26] 'Control of Asbestos at Work Regulations 1987', SI 1987 No. 2115 (ISBN 0 11 078115 5), 1987, HMSO.

[27] 'Protection of Workers from the Risks Related to Exposure to Asbestos at Work', Directive 83/477/EEC, *Official Journal of the European Communities*, L 263, 24.9.83, p 25, HMSO.

[28] 'Classification, Packaging and Labelling of Dangerous Substances Regulations 1984', SI 1984 No. 1244, (ISBN 0 11 047244 6), 1984, and subsequent amendments, HMSO.

[29] Directive 79/831/EEC: 'Sixth Amendment of Directive 67/548/EEC on the Classification, Packaging and Labelling of Dangerous Substances', *Official Journal of the European Communities*, L 259, 15.10.79, pp 10–28, HMSO.

[30] 'Information Approved for the Classification, Packaging and Labelling of Dangerous Substances for Supply and Conveyance by Road', (Authorised and Approved List), Third Edition (ISBN 0 11 885542 5), 1990, HMSO.

[31] 'Classification and Labelling of Substances Dangerous for Supply. Notification of New Substances Regulations 1982. Classification, Packaging and Labelling of Dangerous Substances Regulations 1984, 1988', COP 22, (ISBN 0 11 883990 X) 1988, HMSO.

[32] 'Classification, Packaging and Labelling of Dangerous Preparations', Directive 88/379/EEC, *Official Journal of the European Communities*, L 187, 16.7.88, pp 14–30, HMSO.

[33] 'Health and Safety (First Aid) Regulations 1981', SI 1981 No. 917, (ISBN 0 11 016917 4), 1981, HMSO.

[34] 'First Aid at Work', Approved Code of Practice, COP 42 (ISBN 0 11 885536 0), 1990, HMSO.

[35] 'Reporting of Injuries, Diseases and Dangerous Occurrences Regulations 1985'

(RIDDOR), SI 1985 No. 2023, (ISBN 0 11 058023 0), 1985, HMSO.
[36] 'A Guide to the Reporting of Injuries, Diseases and Dangerous Occurrences Regulations 1985', HSE Guidance Booklet HS(R)23 (ISBN 0 11 883858 X), 1986, HMSO.
[37] Sixth Amendment of the 1967 CPL Directive 67/548/EEC', Directive 79/831/EEC, *Official Journal of the European Communities*, L 259, 15.10.79, pp 10–28, HMSO.
[38] 'Test Methods for Physico-Chemical, Toxicological and Ecotoxicological Properties of Dangerous Substances', Annex to EC Directive 84/449/EEC (the 'Sixth Adaptation'), *Official Journal of the European Communities*, L 251, 19.9.84, pp 1–223, HMSO.
[39] 'Test Methods for Toxicological and Ecotoxicological Properties of New Substances', Annex to EC Directive 87/302/EEC (the 'Ninth Adaptation'), *Official Journal of the European Communities*, L 133, 30.5.88, HMSO.
[40] 'Guide to the Notification of New Substances Regulations', HSE Guidance Booklet HS(R)14 Rev (ISBN 0 11 885454 2), 1988, HMSO.
[41] 'European List of Notified (New) Chemical Substances', *Official Journal of the European Communities*, C 139, 29.5.91, (ISBN 0 11 968831), HMSO.

2

The international regulatory scene

Dr J. R. Jackson

SUMMARY

Many managers and health professionals are employed by multinational companies and may need an understanding of the regulatory environment of the country of their headquarters to understand corporate standards and regulations. Those who work for British companies may be using the standards of other countries or selling into them and need an understanding for these purposes too. Increasingly, UK legislation is required to follow EC directives, and an understanding of EC legislation and regulatory processes is necessary for an understanding of the UK regulatory scene.

There are fundamental differences between the legal systems of different countries; for instance, those whose laws derive from the Napoleonic Code Civile are very different from the British legal system. An understanding of the historical and philosophical differences is necessary for an understanding of the laws written under them.

This chapter includes outlines of occupational health legislation and regulatory systems in some developed countries and in the European Community.

Already, many British companies operating internationally react to EC directives before they are implemented nationally and expend just as much effort contributing to the development of EC legislation as to UK legislation. There is a natural desire amongst international companies for a convergence of both the detail and approach of health protection legislation. It will be difficult for developing countries to accept the common positions of the developed countries and the sophistication of these may be inappropriate in any case. We must expect convergence in developed countries and the 're-invention of the wheel' and diversity from the expression of individuality in developing countries.

1. INTRODUCTION

For many companies operating internationally, differences in national legislation are an unwelcome factor in the environment in which they operate, particularly when differences occur between different legislative fields within the same country. An example of the latter is in the various classifications of chemicals for transport, use, etc.

There have been some welcome initiatives, in recent years, particularly in the field of chemical safety, to bring about international harmonization. It is particularly noteworthy that there is agreement within OECD countries on protocols for toxicological testing and procedures for updating these protocols in the light of technical progress. However, it remains the case that other protocols may also have to be satisfied in OECD member countries, such as the USA, if data submissions are to be readily accepted. Recently, the European Community, the US Chemical Manufacturers' Association and others, meeting under the auspices of the International Council of Chemical Associations (ICCA), have agreed the section headings of a 'World-Wide Safety Data Sheet' for chemical products.

This has been incorporated into European Community legislation as a directive (91/155/EEC), initially applicable to dangerous preparations and, subsequently, to dangerous substances by the 'Seventh Amendment' (92/32/EEC). However, this format is different from that being used by the International Programme on Chemical Safety (IPCS) in its International Chemical Safety Cards (ICSCs), though their intended functions are somewhat different.

Regulatory contexts

Not only are there areas where such welcome initiatives have incomplete penetration but the different national regulatory contexts in which they are set gives the same legislation a different significance in different states. For example, in the USA, safety data sheets are regarded as a means of communicating hazards from the employer to his employees as well as from the supplier to his customers. They sometimes appear to be designed to protect the supplier against charges of withholding information rather than to give clear, informative advice to the customer's management. A British safety data sheet which conforms to the ICCA format can thus appear quite unacceptable to an American parent or subsidiary company.

2. REGULATORY SYSTEMS

Here again, there are major differences. In the UK, as has been explained in Chapter 1, there are four levels of health and safety legislation.† Such a situation can be difficult for many mainland Europeans to understand so that the European Commission has called into question the adequacy of the use by the UK of approved codes of practice in the implementation of EC directives, such as the asbestos directive (83/477/EEC).

---

† Act of Parliament (approved by Parliament), Regulations (approved by a Minister/Secretary of State under authority given to her/him by an Act of Parliament), approved codes of practice (approved by the Health and Safety Commission (HSC)) and guidance notes (also approved by the HSC).

In the USA, standards adopted by the Occupational Safety and Health Administration usually have a single mandatory status and can be challenged in the US courts if a petition is filed within 60 days of issue of the standard. A 'temporary variance' from a standard may be sought by an employer if he cannot meet the compliance date because of shortages of materials, equipment or professional or technical personnel. A 'permanent variance' may be requested—and may be granted —if the employer can demonstrate that his facilities or methods of operation protect his employees 'at least as effectively' as those required by OSHA. This provision applies the same concepts as those underlying the UK 'approved code of practice' system.

The regulatory systems in European Community member states differ considerably owing to their different historical evolution. Two important factors are involved here. One is the extent to which the regulatory system is derived from the *Code Napoléon* and the other is the extent to which occupational disability insurance schemes can determine rules which must be obeyed by employers against financial sanctions. Examples of the latter can be found in France, Germany and Luxembourg. In Greece, Italy and Luxembourg, provisions for the safety and health of workers are written into the national constitutions.

As in the UK and in most developed countries, early health and safety legislation principally covered issues such as general cleanliness, personal hygiene, machinery guarding, control of certain occupational diseases (such as white phosphorus and lead poisoning) and particularly hazardous environments (such as mines). Early general legislation (where it existed), tended to enable the production of more specific legislation. Such legislation appeared in Luxembourg and Greece in the 1920s and in Belgium in the 1950s. In the 1970s, in addition to the USA adopting its Occupational Safety and Health Act and the UK its Health and Safety at Work etc. Act, major legislation was enacted in Germany, Italy, Denmark, France, the Netherlands and Ireland.

In so far as generalization is possible, mainland European legislatory requirements remain fairly specific under over-riding general duties which are set out in a very broad-brush manner. There has been a tendency to warn about hazards (or ban them) and to institute health surveillance to detect their occurrence in individuals rather than to identify, assess and control the hazards.

In many countries, large systems and organizations (such as occupational insurance houses and occupational health services) have evolved to meet regulatory requirements. Such institutions tend to support systems which ensure their continuation. Apart from protection of vested interests, there is great inertia in the machinery surrounding regulatory compliance which makes countries reluctant to harmonize their regulations with those of other nations. In Britain, institutions which have supported the independent professionalism of occupational health and safety practitioners, in an environment where they have been able to exercise considerable judgement, have developed their own inertia. This may not take fully into account the fact that the vast majority of workers do not work in industries blessed with these expert practitioners and have therefore not been able to benefit from them.

As an illustration of different European legislative systems, a detailed consideration is given of the Belgian approach. This entails a description of several aspects which are common to other European states; some of these are indicated, where helpful. In addition to the particular aspects of its occupational health legislation, Belgium also provides examples of the type of constitutional complications which exist in many countries as a result of having three regions (Flanders, Wallonia and the Brussels conurbation), two communities (French-speaking and Flemish) and eleven provinces, each of which has a degree of political independence. Comparable situations of regions (Länder) exist in Germany and linguistic groups (the Catalan-speaking population) in Spain, for example.

3. BELGIAN OCCUPATIONAL HEALTH LEGISLATION

The basic laws on occupational health in Belgium are listed in Table 1 (Environmental Resources Limited, 1985). Virtually all the legislation derived from these are contained in the General Regulations for the Protection of Labour (RGPT).† The regulations of the RGPT apply to most workplaces and employees with a contract of employment (including those employed in the public services). The RGPT is a compilation of legislation from different laws and royal decrees and is arranged in five 'Titles'.

The first Title deals with the planning requirements of classified (as dangerous, unhealthy or annoying) establishments. The more dangerous or annoying (Class I) require permission at the provincial level before they can be built, altered or moved. Class II establishments require permission at the level of the commune, approximately equivalent to a district council in the UK. Classified establishments are supervised by the authorities and regularly inspected by the Technical and Medical Inspectorates and the Classified Establishments Police.

Title II contains general provisions on safety at work and special provisions for young workers; it is divided into four chapters. Chapter 1 deals with basic safety and accident prevention, Chapter 2 with ventilation, lighting, temperature and cleanliness and specifies requirements for sanitary, washing and bathing conditions, etc. Chapter 3 relates particularly to workers' health and Chapter 4 to the employment of people under the age of 18.

Title III deals with particular types of equipment (Chapter 1), particular industries —including the chemical industry—(Chapter 2), particular substances and preparations (Chapter 3), fibrous materials (Chapter 4) and vinyl chloride monomer (Chapter 5). Chapters 3, 4 and 5 have resulted from European Community directives (see below).

Title IV contains old provisions dealing with steam machinery and Title V contains provisions establishing statutory organizations, committees and advisory commissions.

3.1 Governmental control and administration

The Ministry of Employment and Labour is the government department mainly responsible for occupational health and safety. This is divided into two sections

---

† Règlement Général pour la Protection du Travail – RGPT (French)/Algemeen Reglement voor de Arbeidsbescherming – ARAB (Dutch). The texts, though similar, are not identical in translation and reflect and reflect the minor differences in legislation for the two communities.

Table 1. Basic laws on occupational health in Belgium (after ERL, 1985)

Law of 30 August 1919 forbidding the manufacture, import, sale or storage for sale of matches containing white phosphorus.

Royal Decree of 15 September 1919 co-ordinating various legislation on mines and quarries.

Law of 24 February 1921 concerning trade in poisonous, soporific, narcotic, disinfectant or antiseptic substances

Law of 30 March 1926 on the use of ceruse and other white lead pigments.

Law of 6 July 1949 concerning accommodation in industrial, agricultural and commercial undertakings.

Law of 10 June 1952 concerning the health and safety of workers and the healthiness of the workplace.

Law of 11 July 1961 concerning the sale and use of dangerous machines, material and apparatus.

Law of 30 June 1971 relating to administrative fines for infringements of certain social laws.

Law of 16 November 1972 concerning the inspection of work.

Law of 28 December 1977 guaranteeing the protection of industrial doctors.

Law of 5 May 1985 relating to the inspection of dangerous, unhealthy and inconvenient establishments, and supervision of steam engines and boilers.

corresponding to the two communities. There are two main administrative divisions, one covering safety and one covering hygiene and medicine. The Safety Administration is, in fact, responsible for most aspects of occupational health and safety regulations and administers the Technical Inspectorate and the Classified Establishments Police. The Administration for Hygiene and Medicine at Work is responsible for the promotion of hygiene at work and of workplaces, research into the aetiology, pathogenesis and prevention of occupational diseases, investigation of nuisances and the control and supervision of occupational medical services. This last duty is discharged through a tripartite commission. It administers the medical inspectorate (staffed by occupational physicians and nurses) and a laboratory of industrial hygiene.

The Government's principal advisory body is the Higher Council for Safety, Hygiene and Improvement of the Workplace‡ to which all proposed legislation is referred and which may suggest new legislation. It is chaired by the Director-General of the Administration of Safety at Work and includes representatives of Government, employers and employees as well as permanent experts. It has a number of working parties, including one on occupational medicine and health.

Title II provisions
In their current form, the regulations are a result of the implemention of the general provisions of the first 'Framework' directive of the EC (80/1107/EEC) and antecedent legislation.

Exposure to dangerous substances is controlled by

— *Prohibition* (of the use of specified carcinogens, polyhalogenated biphenyls, certain solvents in specified circumstances and the use of beryllium compounds in specified applications).
— *Restriction* (of the use of a list of substances for which permission must be obtained and is given only under certain circumstances and for a limited period).
— *Limitations* (of the numbers of people exposed) and *supervision* (by a qualified chemist) in the indoor use of dangerous substances and preparations.
— *Exclusion* of critical groups (young people, pregnant women and people for whom exposure to harmful substances is certified as 'contra-indicated' by an Occupational Medical Service Doctor) from exposure to chemicals on relevant lists.
— *Limit values* are statutory for substances (e.g. lead or asbestos) covered by EC directives. The Belgian law specifies the Threshold Limit Values published by the American Conference of Governmental Industrial Hygienists (ACGIH TLV^{R}s) as having advisory status. There is also considerable reliance on values derived from the German and Dutch occupational exposure limit lists. Statutory health surveillance requires monitoring of biological and health effects, in many instances. For this, the Ministry of Employment and Labour publishes a list of advisory biological norms, the exceeding of which indicates to the physician the need for rigorous health surveillance. Decisions regarding fitness for further exposure are made on general clinical grounds. (Exceeding the norm is only a contributory factor here.)
— *Ventilation or enclosure* to prevent release into the workplace atmosphere of airborne substances. Where permission is required for the use of harmful substances, primary engineering and other controls may be stipulated in the permit. Where control of emissions at source is not possible, there should be general ventilation and continuing monitoring of atmospheric quality with a view to keeping pollution as low as possible.

‡Conseil Superieure de Sécurité, d'Hygiene et d'Embellissement des Lieux du Travail

The inspectorates have the right to prescribe safe working practices for procedures not covered by specific legislation. Occupational doctors have a duty to examine manufacturing processes and methods of work and propose measures to reduce risks to workers' health.

3.2 Services for occupational health and safety in the workplace
For enterprises with more than 50 employees, the employer must establish a Service of Safety, Hygiene and Improvement in the Workplace. He must appoint a Head of Service, the appointment of whom must be ratified by the Committee of Safety, Hygiene and Improvement in the Workplace. An occupational medical service must be provided. This may either be operated by the company or be an approved inter-company service. In either case, the doctors concerned must be qualified in occupational medicine. The arrangements must be approved by the Committee of Safety, Hygiene and Improvement in the Workplace.

3.3 Exposure records
Exposure records are required for certain substances (i.e., vinyl chloride monomer, asbestos and lead) which are covered by EC directives. The maintenance of the lists of posts having a risk of occupational disease (see below) and their occupants constitutes a qualitative exposure record.

Posts having a risk of occupational diseases
These are identified according to the exposures which may be encountered in them. These are listed in Title II, Chapter 3 in five groups. The list also specifies the particular medical examinations required and their frequency. If the agent is present in the workplace, the level of exposure is not taken into account.

Group 1 consists of jobs with exposure to some 30 broad chemical groups with systemic toxicity; one or more of these are practically unavoidable in the chemical industry. Exposure is usually defined in terms of the duration of exposure each year.

Groups 2 and 3 are jobs involving exposure to physical (ionizing and non-ionizing radiation, noise, vibration, lasers, ultra- and infra-sound) and infective agents. Group 5 are fibres and dusts causing disease of the lungs by inhalation and Group 4 is a miscellany of other occupational exposures.

Employers, in conjunction with the occupational doctor and the Committee for Safety, Hygiene and Improvement of the Workplace (a statutory requirement for all places of employment with more than 50 employees) are required to identify all posts in which employees are at risk of occupational disease.

3.4 Medical examinations
Pre-employment medical examinations are required for:

— all people under the age of 21,
— all people recruited to a post where there is a risk of occupational disease (see previous section),
— those with responsibility for the operation of dangerous machinery or equipment,

- those whose work requires them to have vaccinations,
- food handlers, and
- some handicapped people.

General medical examinations are required for:

- all those for whom pre-employment medical examinations are required (normally annually but at 6-monthly intervals for those under 18),
- pregnant women and
- those who have been absent from work because of illness, injury or childbirth for more than 4 weeks.

Special medical examinations are required for employees who are in posts where there is a risk of occupational disease. Very often, the examination specified is not precisely defined (such as 'respiratory function tests', 'liver function tests', etc).

The format and contents of medical records kept by occupational physicians are specified in detail.

Information on an individual employee provided to the employer is confined to a statement as to his or her fitness to continue work. Anonymized statistical reports can be supplied and a report summarizing the number of examinations undertaken, the number of abnormalities identified, etc., must be submitted annually to the employer and the supervising authority.

3.5 Compensation for occupational diseases

A list of compensatable occupational diseases was first published in the Royal Decree of 3 July 1970. An occupational physician must certify that a disease is present and advise the employee on the procedure for making a claim from the Fund for Occupational Diseases. The claim is considered by a local committee which includes employers' and employees' representatives and experts. Compensation is fairly generous and litigation by the employee against his employer, in such cases, is rare.

3.6 Conclusions

While the Belgian system of occupational health law has some irrational and bureaucratic aspects, particularly in relation to medical surveillance, it has many internal quality control features and disciplines which are absent in the less strictly controlled practices in the UK.

However, as Total Quality Management (TQM) is applied to occupational health practices in the UK, they will move towards Belgian practices in terms of formality. It is to be hoped that other influences on the Belgian legislation will result there in more judgement as to when health surveillance is needed and more control of its relevance and relationship to exposure.

It is noteworthy that, although 'hygiene' is a term found in Belgian legislation, the profession of occupational hygiene is scarcely recognised. The Belgian Society of Occupational Hygiene was formally constituted only in 1991. A similar situation exists in many other European countries.

4. OTHER EC MEMBER STATES

The following highlights the features of particular interest in the legislative systems operating within Britain's main EC trading partners.

4.1 Germany

Germany is a federal republic comprising regional or city states, known as Länder. Länder cannot legislate where there is federal law although they can, and do, adopt legislation which is more specific than federal law. They are responsible for the enforcement of federal law. The majority of older health and safety legislation adopted by Länder has now been superseded by federal law. The State Insurance Code (Reichsversicheringordnung) of 1911 established the principles of accident insurance in Germany and provides for the establishment of Employers' Associations for Accident Insurance (Berufsgenossenschaften, BGs). There are some 35 BGs; in some cases, such as for the building industry, there are several regional BGs. They are co-ordinated by the Central Organization of Industrial Employers' Organizations.

Under the code, the BGs have to make provision, by all suitable means, for the prevention of industrial accidents and for efficient first aid services. BGs have met this requirement by promulgating accident prevention regulations. The Order (of 1980) on Dangerous Substances in the Workplace requires employers to comply with these regulations where they concern dangerous substances. Other regulations may be given the force of law by approval and signature of the Federal Minister for Labour and Social Affairs. Regulations are formulated by commissions composed of representatives of the responsible BG and other BGs, nominees of the federal ministry, representatives of employers and employees, representatives of producers and users of materials subject to regulation and relevant experts. In addition to regulations, BGs can publish guidelines, safety rules, principles and notices.

The Länder have inspection boards which are responsible for ensuring that federal law is implemented and the BGs also have technical inspectorates.

Germany also has a Federal Institute for Industrial Safety and Accident Prevention (Bundesminister für Arbeitsschutz und Unfallforschung). This organization undertakes research, safety testing, data collection and dissemination and training on behalf of the Minister for Labour and Social Affairs. The Deutsche Forschungs-gemeinschaft (German Research Society) has the Commission for Investigation of Harmful Substances in the Workplace (the 'MAK Commission') which establishes occupational exposure limits—the MAK† and TRK‡ values. Otherwise, the principles of legislation are fairly standard.

4.2 Netherlands

Within the Ministry of Social Affairs and Employment exists the Directorate-General of Labour (which is not dissimilar in its function to the UK Health and Safety Commission). It incorporates the Labour Inspectorate, which has some similarities to the Health and Safety Executive in the UK. The main piece of contemporary legislation is the Working Environment—or 'Arbo' Act (Arbeids Omstandingheden-

---

†MAK = Maximale Arbeitsplatz Konzentrationen (Maximum Concentration Values in the Workplace).
‡TRK = Technische Richt Konzentrationen (Technical Exposure Limits).

wet) of 1980. When fully implemented, this will take over the functions of older, more specific legislation, such as the Safety Act of 1934.

4.3 France
The main source of legislation in France is the Labour Code (Book 2, Parts III and IV). Apart from a decree (of 1913) introducing rules concerning general hygiene and safety measures, the majority of legislation dates from the mid-1970s, when occupational health and safety legislation burgeoned worldwide.

Other relevant codes are the Social Security Code and the Public Health Codes. The Labour Code, apart from general and specific legislation, required employers to organize occupational health services and established a labour inspectorate. The Social Security Code defines occupational diseases and established 16 Regional Sickness Insurance Funds (Caisses Régionales d'Assurance des Malades—CRAM) which have duties to research, inspect and elaborate 'technical rules' as well as administer accident insurance and disability pensions.

4.4 Italy
The legislative system in Italy is complex to those unfamiliar with it because it comprises a wide variety of legal instruments, such as the constitution itself, the penal and civil codes, national and regional laws, presidential and ministerial decrees, contracts, agreements and circulars indicating the expectations or intentions of regulatory authorities. It is still undergoing a process of change.

5. EUROPEAN COMMUNITY LEGISLATION

Within the European Community, there is the clear objective of harmonizing minimum standards for occupational health and safety in the member states.

However, while harmonization of minimum standards may serve to raise standards where they are weakest, they do not achieve uniformity since, in the majority of cases, member states are allowed to establish or maintain pre-existing standards which are more rigorous.

Table 2 sets out some early relevant initiatives related to occupational health in the European Communities (originally, three—Coal and Steel, Economic and Euratom), prior to the Single European Act (SEA, effective 1st July 1987).

5.1 European Community institutions
The institutions involved in European Community legislation are discussed below.

The Commission of the European Communities (CEC). This is the 'civil service' of the EC. Strictly, it consists solely of 17 commissioners but each commissioner has his or her own staff ('cabinet') and has special functional responsibilities. Many of these are discharged by one or more of the 23 Directorates General (DGs). Of particular relevance here are the following DGs:

— *DG III* is responsible for the internal market and industrial affairs including chemicals, plastics and rubber (Unit C 4). This unit is concerned with legislation on formulated chemical preparations.

- *DG V* is responsible for employment, industrial relations and social affairs including health and safety (Unit E 2—Industrial Medicine and Hygiene, Unit E 3—Safety at Work).
- *DG XI* is responsible for the environment, nuclear safety and civil protection, including environmental control of products, industrial installations and biotechnology (Unit A 2), emissions from industrial installations and products (Unit A 3) and waste management policy (Unit A 4). Unit A 2 is particularly concerned with legislation associated with the notification, classification, packaging and labelling of dangerous substances.

The Commission initiates legislation by making proposals and supervises the application of European legislation by member states. Its proposals have to be submitted to other statutory Community institutions for an opinion and finally have to be approved by the Council of Ministers or a technical progress committee, on which each member state is represented.

The Council of Ministers is a meeting of ministers of all member states. On different occasions, for different purposes, ministers with different responsibilities will constitute the Council. Thus, on one occasion, it may be formed by, say, agriculture ministers, on another by labour or employment ministers, although foreign ministers are those most commonly involved. Council votes on the issues before it. For some types of issue, since the Single European Act came into effect, an issue may be approved by qualified majority voting (QMV) in which member states have a number of votes related to their population. The rules are such that no single state has a veto. Each member state, in turn, takes the presidency of the Council.

Technical Progress Committees consist of representatives of member states at a sub-ministerial level. Their role is to consider, and usually to approve, adaptations to technical progress of Council directives or to approve Commission directives, which are usually of a technical nature.

The Committee of Permanent Representatives (COREPER) comprises the heads (or their deputies) of the permanent delegations of member states to the European Community. Member states have permanent delegations of which the members, *inter alia*, undertake the preparatory work for the meetings of Council.

The European Parliament (EP) is a directly-elected body and is becoming increasingly powerful and effective. Whilst technically it is only consulted in the process of formulating Community legislation, it is making increasingly effective inputs and has the ultimate sanction of approval of the budget. It meets in plenary sessions but its most important work is undertaken within its committees. Major political groupings have developed in which members affiliated to national political parties with broadly similar views work together.

Economic and Social Committee (ESC) is a body of members who are nominated as individuals representing particular interests. However, they are not allowed to be bound by any instructions from governments or other organizations. They have to be consulted and can express an opinion in some stages of the regulatory process and, on their own initiative, on any aspect of Community activity.

The European Court of Justice, in Luxembourg, is concerned only with interpreting,

Table 2. Early legislative instruments of the European Communities

Year(s)	Instrument
1956–57	Council decisions set up the Mines Safety Commission
1967 and later amendments	Council directive on the Classification, Packaging and Labelling of Dangerous Substances (67/548/EEC)
1974	Council decision to establish an Advisory Committee on Safety, Hygiene and Health Protection at Work
1974	Council resolution made provision for an Action Programme relating to Health and Safety at the Workplace
1975	Council directive on the Co-ordination of provisions in respect of .. Doctors (75/363/EEC). Advisory Committee on Medical Training established
1977-79	Safety signs directives (77/576/EEC and 79/640/EEC)
1978	Council Directive on the Protection of the Health of Workers exposed to Vinyl Chloride Monomer (78/610/EEC)
1980	Council Directive .. laying down Standards for Health Protection of the General Public and Workers against the Dangers of Ionising Radiation (80/836/Euratom)
1980	Council Directive on the Protection of Workers from the Risks of Exposure to Chemical, Physical and Biological Agents at Work — first 'Framework Directive' (80/1107/EEC)
1982	'Seveso' Directive (82/501/EEC)
1982	Council Directive on the Protection of Workers from the Risks related to Exposure to Metallic Lead and its Ionic Compounds at Work (82/605/EEC)
1983	Council Directive on the Protection of Workers from the Risks of Exposure to Asbestos at Work (83/477/EEC)
1986	Council Directive on the Protection of Workers from the Risks related to Exposure to Noise at Work (86/188/EEC)
1988	Council Directive on the Protection of Workers by the Banning of Certain Specified Agents and/or Certain Work Activities (88/364/EEC)

upholding and enforcing Community law. It should be distinguished from the International Court of Justice in The Hague.

5.2 European legal instruments
In addition to the treaties establishing the Community and admitting new members, and to the Single European Act amending the Treaty of Rome, there are the following secondary legal instruments:

- *Regulations* are binding in their entirety. They are directly applicable in member states without national legislation or administration.
- *Directives* are binding as regards the results to be achieved and member states must establish legislation or administrative provisions to achieve these results.
- *Decisions* are binding on the specified addressee(s) but are not generally applicable.
- *Recommendations, Communications* and *Opinions* are not legally binding; communications are no longer used.

5.3 The legislative process
There are three stages in the approval of a Council directive according to the new cooperation procedure established by the Single European Act:

(1) *Development of commission proposals*

The decision to legislate and the choice of the instrument for legislation is taken by the Commission in consultation with other institutions and member states. The initiative may stem from an action programme which has been developed by the Commission and approved by the Council in the light of the opinions of the Parliament and Economic and Social Committee.

The drafting of the legislation passes through several stages as 'working documents' on which there may be informal discussions. It is at this stage that the proposals are most malleable and at which it is most possible to influence the overall shape and scope of the legislation. At a late stage in this process, with legislation emanating from DG V, the Advisory Committee on Safety, Hygiene and Health Protection at Work is consulted. Formal inputs can be made through the Government, industry and trades union representation on this committee. This stage ends with the publication of Commission proposals (usually for Council directives) in the *Official Journal of the European Communities (OJ)*.

(2) *Co-operation procedure*

Commission proposals are considered by the European Parliament and the Economic and Social Committee, which formulate opinions which are published. At this stage, inputs may be made to the relevant Parliamentary committees and by personal lobbying of members of the Economic and Social Committee.

The Commission takes a view on Parliament's opinions and submits proposals (which may be revised) to the Council, which adopts common positions by qualified majority voting. These may involve further alterations to the Commission proposals.

Common positions are submitted to Parliament. If the Parliament approves a common position or takes no view, the Council can adopt the Commission proposal. If Parliament, by an absolute majority of members, amends the common position, the Commission may revise its proposal in the light of the Parliament's amendments and submit it to the Council.

(3) *Decision*

The Council may adopt Commission proposals by a qualified majority vote. Council adoption of parliamentary amendments, not included in the Commission proposal or other amendments, requires a unanimous decision of Council. If the Parliament had rejected the common position, Council can only act by unanimity.

For Commission directives, proposals are submitted to a technical progress committee after opinions of Parliament and the Economic and Social Committee have been received. If a proposal is adopted by the technical progress committee by a qualified majority vote, it becomes law. If it is not adopted, the Commission must make its proposal as a Council directive and it will follow the procedure outlined above.

5.4 The impact of the Single European Act

The Single European Act (SEA) of 1986, effective on 1st July 1987, impacted health and safety legislation principally through two articles, 100a and 118a.

Article 100a

This article requires that

'1. ... The Council shall, acting by a qualified majority on a proposal from the Commission in cooperation with the European Parliament and after consulting the Economic and Social Committee adopt the measures for the approximation of the provisions laid down by law, regulation or administrative action in member states which have as their object the establishment and functioning of the internal market.

'3. The Commission, in its proposal envisaged in paragraph 1 concerning health, safety, environmental protection and consumer protection, will take as a base a high level of protection.'

Article 118a

This article requires that

'1. Member states shall pay particular attention to encouraging improvements, especially in the working environment, as regards the health and safety of workers, and shall set as their objective the harmonisation of conditions in this area, while maintaining the improvements made.

'2. In order to help achieve the objective laid down in the first paragraph, the Council, acting by a qualified majority on a proposal from the Commission, in cooperation with the European Parliament and after consulting the Economic and Social Committee, shall adopt, by means of directives, minimum requirements for gradual implementation, having regard to the conditions and technical rules obtaining in each of the member states.

'Such directives shall avoid imposing administrative, financial and legal constraints in a way which would hold back the creation and development of small and medium-sized undertakings.

'3. The provisions adopted pursuant to this Article shall not prevent any member state from maintaining or introducing more stringent measures for the protection of working conditions compatible with this Treaty.'

These articles established much more clearly than had previously been the case the duty to legislate on health and safety both in relation to the achievement of the internal market and for their own sake. The removal of the requirement for unanimity for approval has made it much easier to approve directives and has accelerated the production of European Community health and safety legislation.

5.5 Legislation since the Single European Act
The first result was an amendment (88/642/EEC) to the first 'Framework Directive', (80/1107/EEC) establishing the procedure for setting exposure limits (i.e., binding limit values (BLVs) and indicative limit values (ILVs)) and establishing a technical progress committee to approve Commission directives establishing ILVs.

In 1989, Council approved a second 'framework' directive on the Introduction of Measures to Encourage Improvements in the Safety and Health of Workers at Work (89/391/EEC). This directive is much more specific than the 80/1107/EEC directive (which continues in force) regarding the general duties and principles of prevention. In particular, it requires that the employer shall 'evaluate the risks to the safety and health of workers, *inter alia* in the choice of work equipment, the chemical substances or preparations used and the fitting out of work places.' (Article 6.3.a).

Other aspects included are the competence ('capabilities') of workers as regards health and safety, worker consultation over the introduction of new technologies, information and training, the designation of one or more employees (with the necessary capabilities and means and in sufficient number) to carry out activities related to the protection and prevention of occupational risks, the provision of first aid and fire-fighting and evacuation facilities, records of assessments of the risks to safety and health, of occupational accidents and of their investigation.

The implementation date of this directive was 31st December 1992.

Subordinate to the second 'framework' directive are numerous 'daughter' directives. The first seven had been adopted before the end of 1990 and concerned workplaces (89/654/EEC), work equipment (89/655/EEC), personal protective equipment (89/656/EEC), manual handling of loads (90/269/EEC), display screen equipment (90/270/EEC), carcinogens (90/394/EEC) and biological agents (90/679/EEC). The implementation dates for these individual directives lie between 31st December 1992 and November 1993.

In 1990, the Commission issued a recommendation (90/326/EEC) to member states concerning the adoption of a European Schedule of Occupational Diseases. In each case, diseases are defined in terms of known (Annex I) or suspected (Annex II) causative agents. Annex III to the recommendation is a brief summary of the situation in each member state and is illustrative of the classification of occupational diseases used in the Community.

Other directives under consideration in 1991 and 1992 include an amendment to the carcinogens directive—an indication of the instability of European legislation at that time. Moreover, legislation is not always produced in an orderly sequence. For example, in mid-1992, European legislation on chemicals depended on the first framework directive (80/1107/EEC, as amended) and its individual directives, the second framework directive (89/391/EEC) and the carcinogens directive subordinate to it. At the same time, there was informal consultation on a 'consolidation' directive on chemical agents, a revision of the carcinogens directive and two parallel developments of the limit values initiative, dependent on the amendment to 80/1107/EEC and, in anticipation of its adoption, the 'consolidation' directive. Issue tracking and legitimate influence are not easy to manage under such circumstances.

Detailed information on the contents of these directives is not appropriate in a text such as this. The full text of EC directives is usually included in UK consultative documents for implementing legislation and, in any case, it is the national implementation of the legislation which has to be complied with. EC directives can be obtained through libraries, Government bookshops and from the Office for Official Publications of the European Communities, 2, rue Mercier, L-2985, Luxembourg.

5.6 Impact of EC legislation on users of chemicals in the UK

The UK Control of Substances Hazardous to Health Regulations 1988 cover most of the requirements but the scope may require extension to assessment of exposures to substances which are not hazardous to health. There will be some impact from health surveillance requirements which will be particular to users of chemicals but the major impacts will relate to non-chemical hazards in the workplace, work equipment, etc. However, these will not particularly affect users of chemicals.

6. THE UNITED STATES OF AMERICA

In the USA, the major basis for the regulation of occupational health with regard to chemicals rests on the Occupational Safety and Health Act (OSHAct) which was passed by Congress in 1970 'to assure as far as possible every working man and woman in the Nation safe and healthful working conditions and to preserve human resources.'

The OSHAct established the Occupational Health and Safety Administration (OSHA) within the Department of Labor:

— to encourage the reduction of workplace hazards,
— to provide for research in occupational safety and health,
— to establish 'separate but dependent responsibilities and rights' for employers and employees for the achievement of better safety and health conditions,
— to maintain a reporting and record-keeping system,
— establish training programmes,
— to develop mandatory job safety and health standards, and
— to provide for the development, analysis, evaluation and approval of state occupational safety and health programmes.

Coverage extends over all territories under Federal Government jurisdiction and is provided either directly by federal OSHA or through OSHA-approved state programmes. The OSHAct does not cover self-employed persons or workplaces already covered by other federal agencies under other statutes. The OSHAct also established the National Institute for Occupational Safety and Health (NIOSH) as an agency of the Department of Health and Human Services. NIOSH conducts research, provides technical assistance to OSHA and recommends standards for adoption by OSHA. NIOSH may, in the course of its research, make workplace investigations, gather testimony from employers and employees and require employers (at NIOSH's expense) to measure exposures, carry out medical examinations and determine the incidence of occupational illness among their employees.

The process of making an OSHA mandatory standard is admirably open and all relevant information is published in the Federal Register, the official record of Federal Government business.

OSHA first publishes a Notice of Proposed Rulemaking—or even an Advance Notice of Proposed Rulemaking—to solicit information which may be helpful in drafting a proposal. The Notice includes the terms of the proposed Rule. A public hearing can be called at the initiative of OSHA or at the request of parties who make a submission in response to the Notice.

At the end of the consultation period, after any public hearing, the arguments heard and the consideration of them are published in detail. The final proposal and the date when it becomes effective are also published.

It is possible to appeal against the final standard in the courts within 60 days of its publication or ask for variances from the standard if compliance is considered impossible by the relevant date because of shortages of materials, equipment or technical/professional resources, or if it can be proved that the methods of operation provide employee protection at least as effective as that required by OSHA. The public may also petition OSHA to review a standard.

Where circumstances require it, OSHA may set Emergency Temporary Standards until a normally approved standard is available. Criteria for these are that workers may be in grave danger and that a standard is necessary for their protection. An Emergency Temporary Standard constitutes both a Notice of Proposed Rulemaking and serves as a Proposed Permanent Standard.

Standards cover most aspects of health and safety. The majority are published in Title 29 of the Code of Federal Regulations (CFR), Part 1910. Title 29 is republished annually incorporating all amendments which have been published in the Federal Register since the previous edition. Title 29 Parts 1900–1910 can be ordered as '29 CFR 1901.1' from the Superintendent of Documents, Government Printing Office, Washington DC 20402 (Telephone: 202-783-3283).

Part 1904 deals with the recording and reporting of occupational injuries and illnesses. Employers with 11 or more employees must maintain records to enable the Bureau of Labor Statistics (BLS) to undertake an Annual Survey of Occupational Illnesses and Injuries. Certain smaller companies may be required to keep statistics for a particular year. All illnesses and injuries leading to death, loss of one or more days' work, restriction of work or motion, loss of consciousness or any injury requiring transfer to another job or medical treatment in excess of first aid must be recorded on OSHA No. 200, Log and Summary of Occupational Injuries and Illnesses. In each case, a more detailed record must be kept on OSHA No. 101 or equivalent. Each year, selected employers receive form OSHA No. 200S which must be completed from information in OSHA No. 200 and submitted.

Standards related to particular agents are contained in Part 1910. Of particular relevance to health protection in relation to chemicals are items in Table 3 below.

Table 3. OSHA standards on health protection from chemicals

Subpart	Reference 1910.NNNN	Subject
C	.20	Access to employee exposure and medical records
G	.94	Ventilation
H	.101–.111	Hazardous materials (flammable, explosive, compressed materials)
H	.120	Hazardous waste
H	.134	Respiratory protection
K	.151	Medical services and first aid
Z	.1000	Air contaminants
Z	.1001	Asbestos
Z	.1002	Coal tar pitch volatiles
Z	.1003	4-Nitrobiphenyl
Z	.1004	α-Naphthylamine
Z	.1006	Methylchloromethyl ether
Z	.1007	3,3′-Dichlorobenzidine and salts
Z	.1008	bis-Chloromethyl ether
Z	.1009	β-Naphthylamine
Z	.1010	Benzidine
Z	.1011	4-Aminodiphenyl

continues

continued

Z	.1012	Ethyleneimine
Z	.1013	β-Propiolactone
Z	.1014	2-Acetylaminofluorene
Z	.1015	4-Dimethylaminoazobenzene
Z	.1016	N-Nitrosodimethylamine
Z	.1017	Vinyl chloride
Z	.1018	Inorganic arsenic
Z	.1025	Lead
Z	.1028	Benzene
Z	.1029	Coke oven emissions
Z	.1030	Bloodborne Pathogens
Z	.1043	Cotton Dust
Z	.1044	1,2-Dibromo-3-chloropropane
Z	.1045	Acrylonitrile
Z	.1047	Ethylene oxide
Z	.1048	Formaldehyde
Z	.1200	Hazard communication

Agent-specific OSHA standards specify handling precautions, engineering and personal protective measures, personal hygiene measures, facilities required for all these, provision for authorization of employees, maintenance and decontamination procedures, signs, labelling, information and training. They may also specify a Permitted Exposure Limit (PEL), health surveillance requirements and recording and reporting requirements. Where relevant, there may be other provisions or requirements.

The OSHA Hazard Communication Standard, 1910.1200, requires manufacturers and importers of chemicals to assess the hazards of their chemicals and transmit the hazard information to users by means of labels, material safety data sheets, training and access to written records. The information supplied should be comprehensive enough to allow employers to devise appropriate employee protection programmes and give employees the information necessary to protect themselves. It came into force completely in 1986. It requires a carcinogenic hazard to be mentioned if the chemical has been assessed as a carcinogen by the International Agency for Research on Cancer (IARC), the US National Toxicology Program or OSHA. There are detailed criteria for other hazards.

7. CONCLUSIONS

Since the principles of good practice in workplace health protection against chemical hazards are generally agreed internationally and, since most legislation is based on these principles, there is much that is common in health protection legislation internationally.

On the other hand, since the basic legal frameworks and social premises are so different in different countries, the style of regulations themselves and the consultative procedures for setting them differ considerably.

Beyond this, the attitude to the law and the degree of enforcement are very variable. Variation in attitudes is illustrated by such clichés as 'The law is the law' and 'The law was never intended to prevent you doing what is right'.

The effectiveness of harmonization of health and safety legislation in achieving good standards and a 'level playing field' for economic activity is limited by harmonization only applying to minimum standards and by the differences in attitude and enforcement. It is likely that upward convergence of health and safety standards will be achieved by the following four activities parallel to regulation:

(a) Total Quality Management systems, including health and safety management,
(b) development of equipment and functional performance standards on an international basis,
(c) development of professional training and competence standards within the relevant professions and
(d) trans-national movement of health and safety professionals.

REFERENCE

[1] *The Law and Practice Concerning Occupational Health in the Member States of the European Community*, Environmental Resources Ltd, for the Commission of the European Communities, 5 volumes (ISBN 0 86010 626 8/–627 6/–628 4/–794 9/ –795 7), Graham & Trotman, London, 1985.

3

Non-governmental organizations

Dr J. R. Jackson

SUMMARY

Certain non-governmental organizations have been selected because of their relevance to occupational health or chemical safety or both. They are considered under the broad headings of:

— major international organizations,
— international industry-supported organizations,
— national industry organizations, and
— academic and similar institutions.

In each case, sufficient general information is provided about the organization for the reader to appreciate its overall purpose and status. Specific information is given about the organization's contribution to health protection in the workplace with particular reference to chemical hazards. In several cases, the major contribution has been to the dissemination, understanding and interpretation of information about chemical hazards.

Where appropriate in the context of this chapter, general discussions are introduced evaluating the contributions of non-governmental organizations generically.

In general, progress has been made by technical and scientific developments within industry or specialized institutions, (some of which are governmental). These are disseminated by industry-supported organizations and then incorporated into the legislation of advanced countries. Subsequently, they are taken up and promoted by major international organizations.

INTRODUCTION

Non-governmental organizations are considered in this chapter under the following main headings:

— major international organizations,
— international industry-supported organizations,
— national industry organizations and
— academic and similar institutions.

These are discussed below.

1. MAJOR INTERNATIONAL ORGANIZATIONS

Apart from the conventions and recommendations of the International Labour Office, the major contribution of international organizations to health protection in relation to chemicals at work has been in the provision, collection and evaluation of information enabling chemical hazard (and, to a lesser extent, risk) assessments to be carried out.

1.1 The United Nations Organization (UNO)

1.1.1 The International Labour Organization (ILO)
The ILO was founded in 1919 under the League of Nations and was the first organization to be affiliated under the United Nations in 1946.

Its functions include the development and promotion of standards for national labour legislation and practice to protect and improve conditions of work and living. It provides technical assistance in social policy and administration and in manpower utilization and training. It encourages international co-operation and rural industries. It compiles labour statistics and undertakes research on various social problems. Between the world wars, it was concerned mainly with the development of basic labour legislastion and combatting unemployment. Since the end of the Second World War, under-developed countries of the third world have become a significant part of its membership and the emphasis has shifted to human rights and technical assistance.

National membership is associated with tripartite representation from each country's government, employers' and workers' organizations.

The organization is based in Geneva and is a co-sponsor of the International Programme on Chemical Safety (IPCS) (discussed below). It produces a wide range of publications on all aspects of work and employment. Lists of these publications are sent to an extensive mailing list on a regular basis. The ILO also publishes the Encyclopaedia of Occupational Health and Safety.

In addition to these informative publications, the ILO publishers recommendations—which member states are invited formally to accept—and conventions—which members are invited to ratify. Ratification binds the country to the provisions of a convention for at least 10 years. A convention can only be ratified in its entirety and applies to all employment sectors. However, national derogations can be made after suitable consultations.

Each convention is usually accompanied by a recommendation on the same subject which supplements and adds detail to the convention. Occasionally, conventions have been adopted without a corresponding recommendation and vice versa. Of relevance to health protection from the effects of chemicals, those in Table 1 are significant.

Table 1. Key ILO conventions and recommendations

Year	Subject	Convention	Recommendation
1971	Benzene	136	144
1974	Carcinogenic Substances and Agents	139	147
1977	Working Environment (air pollution, noise and vibration)	148	156
1981	Occupational Safety and Health and the Working Environment	155	164
1985	Occupational Health Services	161	171
1986	Safety in the Use of Asbestos	162	172
In preparation	Safety in the Use of Chemicals at Work	–	–

The requirements of Convention 155 should not be burdensome to countries with well-developed occupational health and safety practices so that its ratification would not achieve any great improvement. Any significant improvements would necessarily be burdensome. Although there may be some minor political and industrial relations benefits in ratifying such conventions, governments may see ratification as an unnecessary restriction of their independence of action. For example, the UK Government (which had not ratified Convention 155 at the time of writing) sought the advice of the Health and Safety Commission on ratification and a consultative document on the question was published in 1982. This stated that no changes to current legislation or practice would be required to ratify the convention. This document cited that the only specified difference between the ILO recommendations and then-current UK regulations was that the former required that, when protective clothing (and equipment) was provided by employers, it should be at no cost to workers. Section 9 of the Health and Safety at Work etc. Act only requires that protective equipment provided as a result of statutory requirements need be at no cost to the worker

Since recommendations do not have to be accepted in their entirety, this particular aspect of the recommendation could have been disregarded. The reasons for not ratifying and accepting the recommendations cannot have been for difficulty of compliance.

At the International Labour Conference in Geneva in 1989, discussion took place on a proposed convention on safety in the use of chemicals at work which would apply to all branches of economic activity in which chemicals are used. The requirements for ratifying nations would be that:

(1) a system should be in place for classifying and labelling chemicals with regard to hazards,
(2) a system should be in place for providing safety data sheets for hazardous chemicals and information sheets for non-hazardous chemicals,
(3) suppliers of chemicals classified with regard to hazards should identify the chemicals they supply, assess their properties (on the basis of a search of available information) and supply to employers safety data sheets (or information sheets for non-classified chemicals).
(4) Employers should co-operate with workers (and their representatives) to ensure that chemicals are properly labelled, ensure that empty containers are properly disposed of (in accordance with national laws and practices), maintain a register of hazardous chemicals at the workplace (which should be available to workers), protect workers by:
  — choice of chemicals to minimize risk,
  — choice of safe technology, working systems and practices, engineering control measures, adequate occupational hygiene measures and, where these measures do not eliminate the risk,
  — provide personal protective equipment and clothing (at no cost to the worker).
  Employers should limit exposures to levels necessary for health protection, provide first aid and arrangements to deal with emergencies, information and training and monitor exposures when necessary to safeguard their safety and health.
(5) Workers should co-operate with employers and should take all reasonable steps to minimize the risks to themselves. They should have the right to remove themselves from danger and should have rights to information.

There were also proposals for recommendations on similar lines.

While there is nothing new in these proposals, for most developed countries, the proposed wording is felicitous and succinct and should provide a good basis for the development of legislation in countries where such regulations have not yet been produced.

1.1.2 The World Health Organization (WHO)

The WHO was founded in 1948 as a specialized agency of the United Nations. It inherited specific tasks relating to the prevention and treatment of infectious diseases (from the Health Organization of the League of Nations and the International Office of Public Health in Paris) and was given a very broad brief to promote the attainment of 'the highest level of health' by all peoples. The WHO defined 'health' for itself in terms which are generally well-known, i.e., 'a state of complete physical, mental and social well-being and not merely the absence of disease or infirmity'.

The WHO has a general policy-making body—the World Health Assembly—(which meets annually), an Executive Board of health specialists (elected for 3 years by the assembly) and a secretariat. It has six regional organizations which cover the whole world. The Regional Office for Europe is situated in Copenhagen.

Occupational health, in general, falls within the terms of reference of the WHO and there is activity in Europe to encourage international co-operation in achieving a high standard of training and assessment of competence and certification of specialists. Useful conferences are being held to define the contents and boundaries of the various professional disciplines involved in occupational health (e.g., the WHO/EURO Workshop on the Development of the Occupational Hygienist in Europe, held in Geneva, in July 1991). The main activities of the WHO in relation to occupational exposure to chemicals are achieved in co-operation with the ILO and the United Nations Environment Programme (UNEP) in the International Programme on Chemical Safety (see below).

1.1.3 The United Nations Environment Programme (UNEP)
The principal activity of this organization in relation to protection of health from occupational chemical exposure is the maintenance of the International Registry of Potentially Toxic Chemicals (IRPTC). The IRPTC is a potentially useful database of national legislation on chemicals but, in so far as it relies on voluntary submissions, it is not always up to date or complete.

1.1.4 The International Programme on Chemical Safety (IPCS)
The IPCS was formally launched in April 1980 when a memorandum of understanding was signed between the WHO, the ILO and UNEP. It was established specifically to produce assessments of the risks to human health and the environment from chemicals, thus providing the internationally evaluated scientific basis on which member states can develop their own chemical safety measures. The WHO manages the programme on behalf of the three organizations; day-to-day running of the programme is handled by the Central Unit of the IPCS in Geneva. An Inter-Secretariat Co-ordinating Committee ensures close collaboration between the three cooperating organizations.

While the IPCS is responsible for risk assessment, risk management is the responsibility of WHO programmes, such as those on the control of environmental health hazards, food safety and workers' health. Within the WHO, co-ordination of these activities is achieved through an intra-WHO Co-ordinating Committee on Chemical Safety.

Approximately 31% of the IPCS budget is derived from the WHO and UNEP, 31% from the USA and 38% from the rest of the world. The UK contribution is 8.7% of the total. The IPCS collaborates with various other UN organizations, national participating institutions, national focal points and international participating institutions, including the industry-supported scientific organizations, ECETOC and GIFAP (see below).

The IPCS has the following two overall roles:
(1) to provide a forum for establishing international consensus assessments of the risks to health and the environment of chemicals and
(2) to promote the use nationally of these assessments and strengthen the capabilities of member states to deal with chemical emergencies.

These roles are expressed in the following six objectives:
(1) Carry out and disseminate evaluations of the risk to human health and the environment from exposure to chemicals, mixtures of chemicals or combinations of chemicals and physical and biological agents.
(2) Promote the development, improvement, validation and use of methods for laboratory testing and ecological and epidemiological studies and other methods suitable for the evaluation of health and environmental risks and hazards from chemicals.
(3) Promote technical co-operation with member states, in particular developing countries, to
   (a) facilitate the use of available evaluations of health and environmental risks and hazards from chemicals;
   (b) improve the capabilities of national authorities in conducting their own evaluations of health and environmental risks and hazards from chemicals;
   (c) strengthen infrastructures for safety aspects relating to chemicals—their production, importation, transportation, storage, use and disposal.
(4) Promote effective international cooperation with respect to emergencies and accidents involving chemicals.
(5) Support national programmes for prevention and treatment of poisonings involving chemicals.
(6) Promote training of the required manpower.

The IPCS produces documents. The first type are called Environmental Health Criteria (EHC) documents and these numbered some 120 at the end of 1991. The majority are concerned with risk assessments for specific chemicals or groups of chemicals but some are concerned with the methodology of risk assessment, such as EHC 119, published jointly by the IPCS and the Commission of the European Communities, and entitled, 'Principles and Methods for the Assessment of Nephrotoxicity Associated with Exposure to Chemicals'. EHCs on individual chemicals are intended to be comprehensive critical reviews of the literature giving sufficient detail of the studies reported to enable the utility of the studies and their contribution to the evaluation to be assessed by readers who are expected to be technical specialists in the relevant sciences.

Simpler documents, without environmental data, are now being published as Health and Safety Guides (HSGs). These are designed to help employers establish processes and work practices which will avoid risks from chemicals.

A large number of International Chemical Safety Cards (ICSCs) have also been produced. These contain essential information clearly expressed to enable workers to understand the hazards and risks of the chemicals with which they work.

The IPCS has also produced more than 70 Pesticide Information Sheets. Most recently, the IPCS has been producing Poisons Information Monographs which are designed to assist poisons centres and clinical toxicologists in the diagnosis and management of cases of human poisoning.

Though not connected with health protection from occupational exposure to chemicals, the IPCS co-operates with the UN Food and Agriculture Organization (FAO) in developing Acceptable Daily Intakes (ADIs), Tolerable Daily Intakes and Maximum Residue Levels for food additives, pesticides, veterinary drugs and contaminants in foods. The relevant joint committees are the Joint WHO/FAO Expert Committee on Food Additives (JECFA) and the Joint Meeting on Pesticide Residues (JMPR).

1.1.5 The International Agency for Research on Cancer (IARC)

The IARC is an agency of the World Health Organization and has its headquarters in Lyon, France. It was founded in 1965. Its aims are to promote international collaboration in cancer research and, in conjunction with other organizations, to stimulate and support research into the problems of cancer. It has a Governing Council consisting of one representative of each member country and the Director-General of the WHO; this appoints a Scientific Council of 12 scientists.

Its major activities are research in the epidemiology, aetiology and pathogenesis of cancer. It carries out reviews of published data on selected chemicals and classifies them as regards their carcinogenicity. It also publishes state-of-the-art monographs on techniques, the scope of which is illustrated by the following list:

— Molecular and Cellular Aspects of Carcinogen Screening Tests,
— Cancer Mortality by Occupation and Social Class 1851–1971,
— Statistical Methods in Cancer Research (2 volumes) [one of the most definitive texts on epidemiological method available],
— Cancer Incidence in Five Continents (four volumes),
— Age-Related Factors in Carcinogenesis,
— Interpretation of Negative Epidemiological Evidence for Carcinogenicity,
— Serial Directories on On-going Research in Human Cancer and
— Monitoring Human Exposure to Carcinogenic and Mutagenic Agents.

Undoubtedly, the aspect of its work which is best known is its classification system for carcinogens and its monographs which summarize relevant published data on important chemicals. These monographs also contain conclusions on the carcinogenicity of the chemicals reviewed.

The first classification system was established in 1971 and has since been reviewed several times; a major revision of the system was concluded in 1988. In the current system, the groups are defined as follows:–

Group 1: The agent is carcinogenic to humans
This category is only used when there is sufficient evidence for carcinogenicity in humans.

Group 2
This category includes agents and exposure circumstances for which, at one extreme, the degree of evidence of carcinogenicity in humans is almost sufficient and, at the other extreme, there are no human data but for which there is experimental evidence of carcinogenicity.

Group 2A: The agent is probably carcinogenic to humans
The exposure circumstance entails exposures that are probably carcinogenic to humans. This category is used when there is *limited* evidence of carcinogenicity in humans or sufficient evidence of carcinogenicity in animals strengthened by supporting evidence from other relevant data.

Group 2B: The agent is possibly carcinogenic to humans
The exposure circumstance involves exposures that are possibly carcinogenic to humans. This category is generally used when there is limited evidence of carcinogenicity in humans in the absence of sufficient data of carcinogenicity in experimental animals. It may also be used when there is inadequate evidence of carcinogenicity in humans or when human data are non-existent but there is sufficient evidence of carcinogenicity in experimental animals. In some instances agents for which there are no or inadequate data in humans and limited evidence in experimental animals may be included in this group if there are supporting data.

Group 3: The agent is not classifiable as regards carcinogenicity
This category is used where agents do not fall into any other group.

Group 4: The agent is probably not carcinogenic to humans
This category is used where there is evidence suggesting lack of carcinogenicity in humans together with some evidence suggesting lack of carcinogenicity in experimental animals.

1.2 The Organization for Economic Co-operation and Development (OECD)
The OECD was established in 1961 as the successor organization to the Organization for European Economic Co-operation (founded in 1948) and has its headquarters in Paris. It is concerned with all aspects of economic and social policy, covering Economic Policy, Energy and Nuclear Energy, Development Co-operation, International Trade, Financial and Fiscal Affairs, Agriculture and Fisheries, Environment, Science, Technology and Industry, Manpower, Social Affairs and Education and Road Research'. It has established an Environment Committee which has a special Chemicals Programme to promote co-operation in controlling the 80 000 chemicals in commerce.

Industry representations to OECD are made through the Business and Industry Advisory Committee (BIAC) which was constituted in 1962 as an independent organization recognized by the OECD as representative of business and industry. Its membership comprises the industrial and employers' organizations of the 24 OECD member countries. BIAC works through specialist committees which correspond to, and work with, OECD directorates. There are committees on chemicals and the environment which are particularly relevant to chemical safety.

Under its Chemicals Programme, the OECD is carrying out three major activities which contribute to health protection from the effects of chemical exposure.

The first is the establishment of a set of toxicological testing protocols and a system of Good Laboratory Practice (GLP) which has undoubtedly contributed to the quality and extent of toxicological testing and hence to our knowledge of the potential health effects of chemical exposure. The original motivation for this was to prevent differences between national protocols used for chemical safety evaluation being a barrier to trade in chemicals. The OECD protocols have broadly been accepted in all OECD member countries. In the European Community, they have been incorporated into the toxicological testing requirements for new chemicals under the 'Sixth Amendment' (79/831/EEC) to the EC Directive on the Classification, Packaging and Labelling of Dangerous Substances (67/548/EEC)—and hence into the legislation of each EC member state. The avoidance of unnecessary duplication of testing to satisfy individual countries' demands has tended to make more data available for the same expenditure. Although the general introduction of GLP has increased costs, the value for money of testing has probably increased and the results are certainly more reliable.

The OECD protocols are kept under review and are adapted from time to time to technical progress.

The second major programme with a potential to enhance health protection in relation to chemicals is the Existing Chemicals Programme. While most OECD countries have introduced notification programmes for new chemicals which require the generation and submission of toxicological data (Pre-Marketing Notification in Europe, Pre-Manufacture Notification in the USA), these make little impact on the tens of thousands of chemicals in wide use and the hundreds of thousands of chemicals known. Of these, certain High Production Volume chemicals have been identified and dossiers on them produced in OECD member countries (with the assistance of manufacturers) which have summarized the information on these chemicals relevant to a hazard assessment.

It has been recognized that economic and practical considerations (such as the availability of laboratory facilities) do not permit the full set of data required for new chemicals to be generated for even the most commonly used existing chemicals. Consequently, a series of 'screening tests' has been developed to indicate the need for formal testing rather than fully to characterize a toxicological effect.

This second stage of the Existing Chemicals Programme consists of the development of Screening Information Data Sets (SIDs), regarded as the minimum information required for each chemical. It is intended that these should be used where formal, full protocol studies (to OECD or similar protocols) are not available. If the results are not indicative of any hazards, no further tests need be undertaken.

However, this programme is still at an early stage and various details are still to be finalized. Even matters of principle were still under discussion in 1991, for example, whether the potential for exposure to the chemical should contribute to determining the SIDs.

The OECD programme is being integrated with programmes of regulatory authorities, such as that of the National Toxicology Program in the USA and the proposed Existing Chemicals Regulation in the European Community. A common, computerized-data submission dossier (HEDSET) between OECD and the EC has been produced.

The third activity concerns efforts to harmonize the various systems used for the classification of chemicals. At present, different classification systems are used for different purposes and in different countries. This gives rise to numerous complications which have prompted countries to seek harmonization—at a position as close as possible to their own systems!

1.3 International Commission on Occupational Health (ICOH) (and its specialist section on the chemical industry, MEDICHEM)

The ICOH (until a few years ago, the Permanent Commission and International Association on Occupational Health (PCIAOH) and before that the International Permanent Commission on Occupational Health) has been in existence for over 80 years. Membership is open to individuals from any branch of occupational health practice who are sponsored by three members in good standing.

The location of the ICOH secretariat changes with the elected Secretary-Treasurer. At the time of publication, it was located in Singapore.

It has 25 scientific committees, each dealing with issues of a particular sector of industry or a particular aspect of occupational health, such as shiftwork. The committee for the chemical industry is known as MEDICHEM. The ICOH organizes a conference every 3 years. MEDICHEM holds a conference annually; every third year, it is held during the course of the ICOH conference. Other sections, such as that on synthetic organic fibres, are also concerned with health protection in relation to chemicals.

Apart from its conferences and newsletters, which are useful, a particular value of the ICOH is as a forum for interchanges between professionals from different fields of occupational health, particularly those from parts of the world with very different occupational health and economic problems. The membership list is a useful starting point for advice to practitioners in international companies wishing to start up an occupational health function, perhaps for a newly acquired facility, in a country of which they have no experience.

1.4 The European Foundation for the Improvement of Living and Working Conditions (the 'Dublin' Foundation)

The 'Dublin' Foundation was established in 1975. Its headquarters is in Dublin but, in 1991, it opened a representative office in Brussels. Its main concern is with the sociological and labour relations aspects of employment. However, it has been proposed (by UNICE in 1992) that its scope should be expanded in the field of occupational health and safety as part of an alternative to the establishment of a European Agency for Safety, Hygiene and Health Protection at Work. Its publications relevant to health protection from chemicals, listed in its 1990 catalogue, include:

- Safety in Hazardous Wastes: Information Booklet No. 2,
- The Impact of Biotechnology on Working Conditions,
- The Impact Impact of Biotechnology on the Environment,
- Shiftwork in the Chemical Industry,
- Shiftwork in the Chemical Industry: Case-Studies of Innovations and
- Safety and Health in the Workplace (nine volumes corresponding to each of the EC member states in 1980).

Its 1990 activities were divided into six areas and information projects. Area 3 was concerned with promoting health and safety and the major activities were:

- Systems of Monitoring Working Conditions related to Health and Safety in the European Community,
- Screening Techniques in Health and Safety at Work,
- The Consideration of Working Conditions in the Management of Industrial Projects,
- Innovative Workplace Action for Health and
- Ageing of the Workforce and Work Organization.

2 INTERNATIONAL INDUSTRY-SUPPORTED ORGANIZATIONS

2.1 European Chemical Industry Council (CEFIC) and other trade associations

The acronym derives from the French version of its former name—Conseil Européen des Fédérations de l'Industrie Chimique. More recently, major chemical companies have been able to become members directly so that the title of the organization has been changed. However, the acronym has been retained as a convenient short name. It was founded in 1972 by the merger of the European Centre of Chemical Manufacturers' Associations and the International Secretariat of Professional Groups in the Chemical Industry of the European Community countries. In addition to dealing with all other aspects affecting the industry, it has a Technical Affairs Group which is supported by specialist committees in health protection, chemical safety, safety of production, transport safety and environmental protection.

It also has over 50 sector groups dealing with particular product groups, some of which also have committees dealing with issues related to health and safety. For example, the Plasticisers Sector Group has a technical committee with a toxicology working group. While the latter is concerned mainly with consumer safety in, for example, food contact applications, data developed by the group are also relevant to occupational health.

CEFIC has published numerous valuable documents (produced by its specialist committees) dealing with such issues as monitoring strategies, occupational epidemiology, guidelines on occupational health management, etc.

The organization is particularly active in regulatory affairs and carefully monitors the activities of the Commission of the European Communities (CEC). It is formally represented on several EC committees, such as the Working Group on Dangerous Substances (which considers the classification of chemicals to be included in Annex I of the Dangerous Substances Directive). The chairpersons of its main health and safety committees periodically meet officials in DG V for informal discussions and to develop co-operation between regulators and industry in processes of regulation. CEFIC produces position papers and briefs to UNICE (l'Union des Confédérations de l'Industrie et des Employeurs d'Europe), the organization of European industry represented officially on tripartite Community committees.

An important role for such organizations is in ensuring that provisions for democracy and consultation in the development of European legislation work effectively. Chains of communication—from UNICE to CEFIC, to national chemical trade associations, to companies, to their health specialists—are rather long and the communication of the views and experience of workers in the field to those who have the opportunity to influence legislation is difficult and tends to be slow.

While CEFIC is the predominant European trade association concerned with the chemical industry, there are other important related organizations. [See section 2.5, below.]

2.2 European Centre for Ecotoxicology and Toxicology of Chemicals (ECETOC)

Formerly the European Chemical Industry Ecology and Toxicology Centre, ECETOC was founded in 1978 as a non-profit-making organization concerned with the scientific aspects of toxicology and environmental toxicology (ecotoxicology). It is governed by a board of directors elected by delegates of member companies at the general assembly. The board determines the overall direction to be taken and appoints the director-general and members of the Scientific Committee.

Through careful restriction of its involvement to strictly scientific aspects and maintaining a consistently high standard of scientific quality in its publications, ECETOC has achieved a reputation for objectivity and scientific independence. This enables industry to apply its vast resources of scientific knowledge and experience to influencing regulatory affairs and practices related to chemical safety, environmental and worker health protection without suspicions of bias or compromise related to commercial considerations.

The major outputs are publications and meetings or conferences. Publications are issued in several series. Joint Assessments of Commodity Chemicals (JACC) reports are reviews similar to the Environmental Health Criteria documents of the IPCS. They have been used as draft EHC documents and some (such as those on potential substitutes for chlorofluorocarbons (CFCs)) have been produced at the request of the IPCS, for which ECETOC is a Collaborating Organization. Monographs produced by ECETOC deal with scientific issues or techniques not related to particular chemicals. Technical reports generally relate to particular toxicological issues or ecotoxicological issues of particular chemicals or experimental techniques.

ECETOC's major contribution to health protection from chemicals is the development and communication of the understanding of the scientific principles underlying hazard identification and risk assessment of chemicals.

ECETOC is usually invited to be represented at expert groups considering EHC documents, at IARC meetings and similar scientific groups. It is seeking to contribute to purely scientific components of regulatory activity—such as the Scientific Experts

Group established by DG V of the CEC to make recommendations for Indicative Limit Values (occupational exposure limits) based on criteria documents.

2.3 The International Council of Chemical Associations (ICCA)

This is a relatively new organization which provides a forum for co-operation between major national and international trade associations in the chemical industry. A recent achievement has been the acceptance of the format for safety data sheets which had been developed by CEFIC and agreed by the Chemical Manufacturers' Association (USA). This has now been incorporated into the EC directive on safety data sheets which applies to dangerous preparations and substances.

The possibility of agreement on a common global format for safety data sheets will make an important contribution to health protection since it will encourage better data communication and will make data more accessible.

2.4 BIBRA Toxicology International

Formerly called the British Industrial Biological Research Association, it was founded about 1965 and is funded by industry and the UK Ministry of Agriculture, Fisheries and Food. As industry has become more international, so has the work of BIBRA, hence the change of name.

Its main roles are to undertake fundamental and contract toxicological research and to provide an information service to its member companies. The monthly bulletin includes summaries of worldwide regulatory developments of concern to the chemical industry and articles of general interest related to chemical safety and toxicology. It has an excellent information department which can help members undertake hazard identification procedures. Its scientists are available, within limits in an advisory capacity, to members and it provides the usual services of any toxicological testing house.

It maintains and publishes toxicological profiles on an increasing number of commonly-used chemicals which present, in an abbreviated format, major information on chemical hazards and their evaluation. It also edits scientific journals and organizes scientific meetings of interest to sub-groups of the chemical industry.

2.5 Sector organizations

The Oil Companies' European Organization for Environmental and Health Protection (CONCAWE) is oriented towards oil and petrochemicals, has a secretariat which includes experienced professionals in relevant disciplines and has produced numerous useful documents, some of which are of general occupational health interest.

The Groupement International des Associations Nationales de Fabricants de Produits Agrochimiques (GIFAP), the International Group of National Associations of Manufacturers of Agricultural Products, has a toxicology committee which makes important contributions to chemical safety related to agrochemicals and has published useful booklets about the safe use of crop protection chemicals, including very practical issues such as the use of protective clothing in hot climates.

The European Association of Metals (EUROMETAUX), the Association of Plastics Manufacturers in Europe (APME), the International Working Group on the Toxicology of Rubber Additives (WTR) and many others have groups paralleling the activities of CEFIC and, in some cases to some degree, ECETOC with regard to their particular viewpoint.

3. NATIONAL INDUSTRY ORGANIZATIONS

3.1 National trade associations representing the chemical industry as a whole

The Chemical Industries' Association (CIA) in the UK has a comprehensive organization dealing with safety, health, environmental and distribution issues in the UK chemical industry. This is managed by the Chemical Industry Safety, Health and Environment Council (CISHEC) on which senior managers from major UK chemical companies are represented. CISHEC is advised by groups dealing with health, safety, environment, distribution and chemical products which, in turn, are served by working groups and task forces. The CIA is very effective in communicating and liaising on issues concerning safety, hygiene and health protection from chemicals in the workplace with the UK Health and Safety Commission and Executive and has produced a large number of valuable guidance documents for use by its members, particularly in relation to the Control of Substances Hazardous to Health (COSHH) Regulations.

Under the auspices of the Responsible Care programme, the CIA is revising some of these documents and has published guidance on safety, health and environmental auditing. It is also developing indicators to be used by the industry to measure its improvement in performance over time.

The CIA is often asked to nominate experts from its member companies or secretariat to be industry representatives on behalf of the CBI, particularly where the chemical industry has a special interest (such as in the Advisory Committee on Toxic Substances (ACTS) and its Working Group for the Assessment of Toxic Chemicals (WATCH) in relation to the setting of UK occupational exposure limits) or special experience (such as in the Industrial Injuries Advisory Council of the Government's Department of Social Security).

3.2 National sector organizations

Many other national, more specialized trade associations in the UK have made important contributions. Two examples must suffice.

The British Rubber Manufacturers' Association (BRMA) has carried out important epidemiological and follow-up work in relation to the occurrence of occupational bladder cancer in the industry.

The British Agrochemicals Association (BAA, an affiliate of the CIA) has Toxicology and Health and Safety Committees specializing in crop-protection chemicals.

Other countries have similar associations representing the mainstream chemical industry and its more specialist sectors. It is not possible within the scope of this chapter to go into these in detail.

4. ACADEMIC AND SIMILAR INSTITUTIONS

4.1 Universities

Universities have always been important because of their research and teaching roles in occupational health. In some countries, the basis for the quality of training rests with the competence and integrity of university departments. Some universities have concentrated on a single aspect of occupational health and, over the years have made considerable contributions. For example, the Catholic University of Louvain has made a great contribution to biological monitoring, an area of particular importance to the chemical industry.

Increasingly, university departments in the UK are having to become essentially self-funding from their consultancy activities. Their collaboration with industry, particularly in the field of epidemiology, has been valuable. Several cases have occurred of co-appointments to full-time positions involving work split part-time between industry and an academic department. Such appointments and consultancy activities are important in ensuring that universities remain relevant in their teaching and research and help to instil scientific rigour into industrial occupational health practice.

Faculty members in universities are also important because they may be independent members or nominees of one or other of the social partners on expert committees involved in the regulatory process. Mutual confidence and respect are necessary if such channels of communication for industrial expertise and data are to be used to the greatest effect in achieving health protection.

4.2 National institutions with a statutory or near-statutory role (but with a significant degree of independence)

In many countries, there are organizations which may be established by legislation (and may be accountable to a government department) but which exert a considerable degree of professional or scientific independence within their charter and contribute to the development of information and strategies for occupational health protection.

To some extent, divisions of the UK Health and Safety Executive fulfil this role. The National Institute for Occupational Safety and Health (NIOSH) in the USA is another important example. In Germany, each main industrial sector has its employment accident insurance fund (Berufsgenossenschaft, usually abbreviated to BG). That for the chemical industry is known as the BG Chemie. These insurance funds undertake research into the causation of accidents and diseases and publish guidance and standards to help the industries they serve to avoid accidents. Compliance with such standards is achieved because employers (rather than the insurance fund) may be held to be liable if the standards are not followed. In Germany, the BG Chemie is undertaking a systematic review of the toxicity of existing chemicals with a view to ensuring that hazard identification is adequate and data gaps are identified.

4.3 National, industry-supported organizations

BIBRA in the UK used to be in this category but has now taken on a more international role (see above). Owing to the resources needed to fund significant organizations, most of these bodies are either international or situated in the USA.

The Chemical Industry Institute of Toxicology (CIIT) is an example of one such industry-funded, scientific organization established in the USA. It undertakes both fundamental and collaborative applied research. It also publishes a periodical, 'CIIT Activities', which reviews key toxicological issues in the context of the CIIT work in the area.

The American Industrial Health Council (AIHC) was established by the American chemical industry and plays a role in the USA similar to that fulfilled by ECETOC in Europe; there is good co-operation between the organizations.

4.4 Professional organizations.

The contribution of professional organizations to the development of occupational health cannot be over-emphasized. The most obvious example is that of the American Conference of Governmental Industrial Hygienists (ACGIH) whose work on occupational exposure limits (TLV^{R}s) is, in a very real sense, the basis of occupational exposure limits world-wide.

The publications lists of most professional organizations give a clear indication of their influence on good practice and competence in their respective disciplines.

Perhaps the most important contribution of all is the role of professional bodies preventing professional isolation. Companies who employ occupational health professionals rarely have more than one of each discipline on each location. Meetings organized by professional bodies help to keep such professionals in touch with developments and colleagues in other companies or organizations (such as academia) whom they may need to approach for advice.

4.5 Standards institutions

Standards institutions, such as the International Standard Organization (ISO), The Comité Européen de Normalisation (CEN), the American National Standards Institute (ANSI), the British Standards Institution (BSI) and the Deutsche Industrie Normung (DIN) have played a role in relation to defining standards for equipment used in occupational health, including respiratory protective equipment. ANSI has published criteria for carcinogenicity of chemicals and CEN is developing standards for monitoring strategies. As Total Quality Management (TQM) principles become adopted by occupational health departments, BS 5750 and ISO 9000 series standards will exert a considerable influence on occupational health practice. Indeed, the diversity in sub-division and training of occupational health professionals in the European Community and the difficulty in defining competence, in terms of a training programme and qualifications, are likely to lead to an increase in the future use of standards for the conduct of individual aspects of occupational health practice in all its branches. This would enable there to be assurance of competence in a particular functional activity, irrespective of the professional title of the person carrying it out.

5. CONCLUSIONS

On the whole, progress in health protection from chemicals, in particular, but also from all other occupational hazards, has taken place in individual practice or in research. The spread of the knowledge or technique recognized as an improvement has almost always occurred fairly effectively before it is taken up in legislation and made a statutory requirement. New knowledge and techniques are not only disseminated by non-governmental organizations, they are validated by them in the process. For this reason, non-governmental organizations have had, have and will continue to have a crucial role in the development of standards of health protection.

4

Occupational health services

Dr C. P. Juniper

SUMMARY

A working definition of Occupational Health (OH) is given, related to current international legislation and followed by an explanation of the relevance of the International Labour Organization (ILO) 1959 Recommendation 112, the 1985 Convention 161 and supporting Recommendation 171 on OH.

Current UK legislation bearing upon the chemical industry is defined. Regulations regarding health, safety, hazard control and reporting of incidents are related to the role of regulatory authorities.

Against this backcloth, the aims and objectives of providing occupational health services (OHSs) are explained together with the preventive aspects of such services. The reasons for, the benefits and creation of, OHSs in industrial undertakings are presented.

How OHSs can be organized and be cost-effective on small, medium or large sites is described with reference to National Health and other medical services in the locality. Indications of OH staffing levels for these sites are provided.

Methods of recruitment, employment and remuneration of occupational health staff are outlined and their relationship with the site management structure discussed.

For the purposes of this chapter, OH (the physical and mental well-being of employees) is distinguished from occupational hygiene (the control of physical, chemical and biological factors in the workplace which may affect the health of employees). It is freely admitted that this distinction is arbitrary as many OHSs also provide hygiene services.

1. A WORKING DEFINITION OF OH

> 'That branch of medical science concerned with the two-way relationship between a person and his or her work, namely the effects of work upon health and of ill-health upon the capacity to work.'

At the present time, there are no legislative requirements in the UK for employers to provide OHSs. This is contrary to the situation in many other countries and most European Community (EC) member states where there is an obligation on employers to do so. In some, the size, complexity and staffing of such services are based on the numbers of employees on site and, in others, note is also taken of the nature of the undertaking.

Disquiet has been expressed about the lack of such an obligation in the UK. As industry does not always have to bear the direct cost of OH, this can give this country a 'competitive edge' financially over other EC states.

However, regulations discussed earlier (Chapter 1) make it incumbent upon employers to provide protection and safe working conditions for their employees. In order to satisfy these objectives, employers need to engage the services of health professionals to a greater or lesser extent.

The International Labour Organization (ILO) in 1959, by means of Recommendation 112 [1], outlined the organization and role of an occupational health service (OHS). This remains the basic international reference on OH. It also laid down the framework for OH which serves as a world-wide basis for this discipline and for professionals working in it.

The 1985 ILO Convention 161 on Occupational Health [2] (together with its supporting Recommendation 171 [3]) is based upon the concept of prevention, rather than cure, and commits states who have ratified it to the provision of occupational health facilities for all in employment. Although the convention is not specific as to how these facilities are to be provided and what they should comprise, the obvious solution is OHSs as provided in many parts of the world.

Article 5 of Convention 161 outlines the functions which are adequate and appropriate to the occupational risks of the undertaking:

(a) identification and assessment of the risks from health hazards in the workplace;
(b) surveillance of factors in the working environment and working practices which may affect workers' health, including sanitary installations, canteens and housing where these facilities are provided by the employer;
(c) advice on planning and organization of work, including the design of the work-place; the choice, maintenance and condition of machinery and other equipment, and substances used in work;
(d) participation in the development of programmes for the improvement of working practices, as well as testing and evaluation of health aspects of new equipment;
(e) advice on occupational health, safety and hygiene, ergonomics, and individual and collective protective equipment;
(f) surveillance of workers' health in relation to work;

(g) promoting the adaptation of work to the worker;
(h) contribution to measures of vocational rehabilitation;
(i) collaboration in providing information, training and education in the fields of occupational health, hygiene and ergonomics;
(j) organizing first aid and emergency treatment;
(k) participation in analysis of occupational accidents and occupational diseases.

There are a number of regulatory requirements in the UK health and safety field which necessitate resources in these areas to be provided by employers.

The Health and Safety at Work etc. Act 1974 (HSW Act) [4] describes the overall responsibilities; the Control of Substances Hazardous to Health (COSHH) Regulations 1988 [5] contain requirements on control; the Reporting of Injuries, Diseases and Dangerous Occurrences Regulations (RIDDOR) 1985 [6] lay down what must be reported to the regulatory authority, namely the Health and Safety Executive (HSE).

The main two branches of the HSE involved with the chemical industry are Her Majesty's Factory Inspectorate (HMFI) and the Employment Medical Advisory Service (EMAS). These give advice as well as monitoring compliance by factories and operating units.

The HSC, in its 'Plan of Work for 1990–91 and Beyond' [7], states that:

> 'Employees in many large firms have access to occupational health services; most employees of smaller companies do not. We shall intensify our encouragement to employers to tackle occupational health problems systematically, both in the course of inspections and by continuing our campaign for the provision ...... of suitable occupational health skills at work'.

These OH skills at work are supplied by OHSs.

2. THE AIM AND OBJECTIVES OF OHSs

The aim of OHSs is:

> 'To provide expert professional support and advice to both management and employees in order to promote occupational health and safety. The service should therefore be primarily, but not exclusively, preventive in nature.'

The objectives of OHSs can be summarized as follows:

(a) prevention of illness or injury caused by conditions in the workplace;
(b) provision of immediate treatment of sickness or injury at the workplace;
(c) placement of employees in a work environment appropriate to their physical and psychological capabilities;
(d) promotion of the general well-being of employees to enable them to work effectively;
(e) provision of advice on OH matters to management and employees;
(f) ensuring that there is compliance with statutory obligations on the health and safety of employees and the environment, both inside and outside the chemical plant.

To achieve these objectives, OH staff must have up-to-date knowledge of plants, processes, products, by-products and all related hazards. They will also have to carry out a range of activities and functions. The scope, diversity and varying size of chemical plants and number of employees on site mean that the following list should serve only as a general indication of what may be required.

3. PREVENTION OF ILLNESS AND INJURY RELATED TO WORK

To do this, OH staff, in general terms, should:

(a) Identify and assess potential hazards to health and safety in the working environment and provide appropriate constructive advice to management and workers about control measures, protective clothing and equipment. Advise on the prevention or control of hazards from factory processes which may affect the public in the neighbouring environment.
(b) Monitor, by medical examinations, health screening or other appropriate methods of surveillance, the health of employees exposed to hazards, before, during and after work. Where appropriate, monitor the working environment or arrange for specialist measurements to be carried out.
(c) Record all episodes of sickness and ill-health at work, injuries due to accidents at work and treatments provided by the OH service and first aiders. Analyse such records to provide relevant statistics to demonstrate trends and problem areas. From these, identify and implement preventive measures.
(d) Advise management on compliance with occupational health and safety legislation.
(e) Advise management on the occupational health aspects of new projects and new technologies.
(f) Consult and liaise with technical and operational management, safety officers, local statutory bodies and national professional organizations concerned with OH.

The following areas also need to be addressed by OH staff:

(g) Promote and maintain good health for all employees by:
   (i) pre-employment or pre-placement health checks;
   (ii) periodic health checks for certain categories of employees;
   (iii) health assessment, after sickness absence, occupational injury or any illness of substantial duration; and
   (iv) health education and other general preventive health measures.
(h) Provide first aid, nursing or medical care by means of:
   (i) OH services, ensuring that immediate, first aid is available;
   (ii) expertise to cope with any special chemical hazards; and
   (iii) ancillary medical services such as X-ray, laboratory, physiotherapy, dental and other services where these can be shown to be cost-effective.

4. ASSESSMENT OF CHEMICAL SITES

The chemical industry can pose potential or real hazards to those working in it. Where risks are perceived and cannot be wholly prevented or adequately controlled, health surveillance is required.

COSHH assessments should reveal the extent of risks and, where these cannot be eliminated, provide the opportunity for arranging the necessary health surveillance. This can be at several 'levels', ranging from an appraisal of sickness and sickness absence, to regular, full health screening and medical monitoring of exposed workers.

In addition to decisions arising from COSHH assessments, a number of statutory obligations may apply. For example, chemical plants which process chemicals such as lead, vinyl chloride monomer or isocyanates are subject to strict occupational health and hygiene regulations which include full health screening.

In 'borderline' instances, where there is doubt about the presence of risk, a low level of health surveillance is appropriate.

COSHH assessments, prevention of ill-health and health surveillance are all part of the tasks carried out by OHSs. Consideration is now given to the provision of these services.

5. WHY PROVIDE OHSs?

Why should employers provide health care at work?

5.1 The law

The range of legal requirements under the HSW Act has been steadily extended by supporting regulations dealing with specific problems. Directives from the European Community are increasing in number and will add to the present statutory requirements. Early warning of these can be helpful so that arrangements for compliance can be planned well in advance. Legally, under the HSW Act, employers must do 'all that is reasonably practicable' to prevent ill-health and risks to their employees. This means that, not only must regulations be obeyed, but employers should go further, by extending surveillance and care into areas where, at present, no legal obligations exist.

5.2 Prevention of ill-health and accidents

Conscientious, caring employers seek to do everything possible to prevent illness and injury to their employees. OHSs have a prime responsibility for, and emphasis on, prevention. A healthy workforce should be more productive and labour turnover lower.

5.3 Reduction of time lost due to illness and injuries

Illness and injuries are costly to employers, particularly in 'slimmed down' modern chemical plants. Steps taken by employers to reduce time lost are potentially beneficial to productivity and competitiveness.

Prompt on-site treatment can, in many cases, prevent or reduce time lost and enhance recovery. It is not unknown for patients to be absent for half a day in an out-patient or casualty department at the local hospital and then to have a few days away from work. Treatment on-site often entails minimal absence from the job.

5.4. Better informed management decisions
Information on the health of potential employees can enable informed decisions to be made about recruitment or placement in jobs. Placing individuals in unsuitable occupations can thereby be prevented and obviate suffering and cost.

Information on hazardous chemicals, processes or by-products, when planning an operation, can ensure that overall costs are contained or reduced. There is no doubt that careful initial planning is less costly than modifying existing plant and dealing with illness arising from it.

5.5 The 'feel good factor'
Many chemical companies subscribe to the concept of the 'responsible and caring employer'. In surveys of what employees see as 'benefits', OHSs consistently appear near the top of the list. Although this is, in itself, insufficient reason for providing OHSs the psychological impact should not be ignored.

6. PROVISION OF AN OHS
There are three main methods of providing OHSs in the chemical industry. The first is to use the services of a local family practitioner, the second is to have a contract with an established, third-party professional service and the third is to employ directly the relevant staff in an occupational health centre (OHC).

As outlined earlier, some countries have legislated for the provision of OHSs by employers. For example, regulations may require a part-time or full-time nurse or part-time doctor to care for 100–500 employees; for 500–1000 employees, one or more full-time nurses and a part-time OH physician; 1000–1500, one or two full-time nurses for each shift and one or more full-time OH physicians, with back-up staff.

In many medium and larger chemical undertakings in the UK, a similar pattern emerges based, not only on numbers, but more importantly on the nature of the business, hazards on the site and the nature of any risks to the employees. More details are provided in Table 1.

Whether these services should be bought in on an 'ad hoc' basis, be provided by annual contract or be an 'in-house' OHS depends on the size of the undertaking, the nature of the work, risk assessment and, not least, the relative costs of each method.

Whichever type of OH cover is envisaged, expert OH advice will be needed to set up a service where none exists.

EMAS, through its local offices, gives overall advice and the secretariat of the Chemical Industries Association may be able to suggest consultants who have a detailed OH knowledge of the chemical industry. Such consultants have the experience which can help management assess the OH requirements of the factory.

Table 1. Levels of OH service

	Level 1	Level 2	Level 3	Level 4
Type of chemical plant	simple products, normal hazards well controlled	multiple products but as Level 1	more complex products, more hazards	complex plant, multiple hazards
Number of employees on site	up to 100	100–200	200–300	300 upwards
Registered Medical Practitioner (Dr)	part-time			
Dr. with Associateship of the Faculty of Occupational Medicine (AFOM)		part-time	part-time or full-time	
Dr. with Membership of the Faculty of Occupational Medicine (MFOM)				full-time
Registered General Nurse (RGN)				full-time or part-time
RGN with Occupational Health Nursing Certificate (OHNC) or Diploma (OHND)	part-time	full-time	full-time nurse in charge	full-time nurse in charge
Enrolled Nurse (EN)			part-time nurse in team	full-time shift cover

When the site has been assessed and the hazards and complexity of the processes evaluated, the potential workload of the OHS should be apparent. Where requirements —statutory or otherwise—dictate regular health surveillance and screening of a significant number of the workforce, and where treatment of ailments or injuries are reasonably frequent, it should be cost-effective to provide an 'in-house' OH service. This applies particularly to larger sites.

7. SETTING UP A SITE OHS

Management should obtain expert advice before embarking on planning and setting up an OHS. Having decided to develop a service, the first step is to draw up a company OH policy, followed by a description of the scope and detail of the service to be provided. This should be based upon the 'Aim and Objectives of OHSs' described in section 2 above. Secondly, and following on from this, job descriptions for the staff should provide content, firm reporting lines and place of each person in the organization. Thirdly, details of the service to be provided and the job descriptions should provide the framework for discussions with the professionals to be employed.

7.1 The occupational health centre
Space is needed for an OH centre which, in the smallest unit, may serve also as the first aid room.

A medium-sized or larger site, particularly where health screening, biological testing and medical investigations are part of the normal workload, needs larger premises. The siting of the health centre, number of rooms, layout and facilities should be central to the plan agreed with the consultant adviser.

7.2 The occupational health nurse

At present, there are two basic kinds of trained nurse, the Enrolled Nurse (EN) or the Registered General Nurse (RGN). The EN (G) is trained as a general nurse after a 2-year course and the RGN after 3 years. Both categories can have further training in OH but to different levels. The first level, the 'Occupational Practice Nurse Award', is open to both ENs and RGNs but the higher level, 'Occupational Health Nursing Certificate' (OHNC) or 'Occupational Health Nursing Diploma' (OHND), is only available to RGNs. These OH nursing qualifications may be obtained by full-time, day-release or modular courses.

An operating unit of medium- to large-size or complexity in the chemical industry requires the nursing leadership of an RGN with OHNC or OHND. A nurse or nurses in the team may have RGN or EN qualifications but should be prepared to obtain the relevant higher degree of training.

The nurse in charge should be of sufficient status and experience to be able to manage the OH team and be in day-to-day charge of the OHS. The same should apply on a smaller- or medium-sized site without additional nurses but with a visiting (part-time) doctor. Further managerial training may need to be provided.

Salary scales are published annually by the Royal College of Nursing (RCN) and should be viewed only as recommendations. OH nurses should understand that each employing company will have its own financial and wage structure and that local employment conditions need to be taken into consideration.

7.3 The occupational physician

There are a number of qualifications which a doctor may obtain in order to be registered with the General Medical Council (GMC) and thereby be allowed to practise in the UK.

In order to specialize in any particular branch of medicine or surgery, acquisition of wide experience and additional qualifications in that speciality are required. Occupational medicine is no exception. Associateship (AFOM), Membership (MFOM) or Fellowship (FFOM) of the Faculty of Occupational Medicine at the Royal College of Physicians (RCP), followed by accreditation as a recognised specialist, is becoming a general requirement. Doctors can start in industry without these specialist qualifications but should seek to acquire the necessary experience and training as soon as possible. They can achieve this training by full- or part-time day-release or a distance-learning course under supervision. Details of courses and examination syllabi are obtainable from the RCP in London.

The British Medical Association publishes annual salary recommendations for full- and part-time physicians working in OH which form the basis for remuneration. The salary and employment conditions of the full-time OH doctor should be within the salary scales and benefits of the company management.

7.4 The part-time doctor

A small site of one hundred or fewer employees can well be served by a local doctor who may be contracted to attend for 1 or 2 hours a week and provide advice over the telephone at other times. On a slightly larger site, attendance may be for two sessions, one at the beginning and another at the end of the week.

Some general practitioners (GPs) work part-time in the chemical industry and provide an excellent service. They are regular attenders on site, understand the processes, have the confidence of the employees and are key members of the management team.

Illness and accidents occur in the general population, whether at home, working in an office or in a chemical factory. It is often difficult to determine whether an illness is work-related and certainly much more so if the doctor is not a regular visitor to the factory. It is essential to understand the precise nature of the employee's job, how he carries it out and what contact he has with chemicals and processes. It is only by doing so that the cause of his condition can be identified.

It is not possible to advise a chemical factory from a distance and without frequent site visits. It is also necessary to have a proper understanding of technical developments and personnel issues in order to give comprehensive advice. The part-time medical adviser should, from time to time, attend safety committee meetings and address the site council on topical OH issues.

Although, in some cases, it may be satisfactory to carry out some screening at the GP's surgery, it is more efficient for the company and worker to be seen and examined on-site and time will thereby be saved. Therefore, it follows that employment of a part-time doctor should be conditional upon regular attendance at the factory.

Some factories pay the GP by means of a retainer (annual sum), with additional fees for each item of service. This method of payment is expensive for the employer and, if the service provided is only to carry out medical examinations at the GP's surgery, it rarely proves satisfactory, or even begins to fulfil present-day, OH needs. It is far better to base remuneration on an assessment of the number of hours required for the doctor to fulfil his commitments. A fee to cover the entire service can then be agreed. Should circumstances or workload change, the hourly commitment and financial arrangements can be reviewed.

7.5 A third party OH service

In some parts of the UK, and located in industrial areas, well-established OH services are provided by experienced staff on a contractual basis to local companies. These exist, for example, in areas such as Greater London, Slough, Manchester, Birmingham, Newcastle and Edinburgh. Some National Health Service hospitals also have OH departments able to do work for third parties.

These organizations are professionally-based, can provide an initial assessment of the factory's OH needs, give an estimate of the cost and then supply the required service.

Whichever type of service is chosen, care is needed to ensure that it is convenient and provides continuity and firm advice to management and employees at reasonable cost.

7.6 The occupational health team

The smallest site may have insufficient workload to employ full-time staff so that a doctor or nurse may cover the OH needs on a sessional basis. However, it must be remembered that the chemical industry is a specialized area in OH. The complexities of modern processes and the variety of raw materials, intermediates and end-products often require a detailed and continually updated knowledge of chemistry, toxicology, physiological responses, biological measurement and epidemiology. The quality of the OH medical team must therefore be of prime importance. The 'core' team should comprise a nurse and doctor who are suitably qualified and experienced together with appropriate back-up staff.

Increase in size and complexity of the OH unit requires an enlarged team. Table 1 indicates the qualified nurses and doctors needed for four levels of service, increasing in effort from Levels 1 to 4.

On sites at Levels 1 and 2, and, in some cases Level 3, the nurse may assume the *managerial* role for the nursing and clerical team and the doctor can function as a visiting consultant. However, the ultimate, *clinical* responsibility rests with the doctor who authorizes and signs 'standing orders' for agreed medical procedures and any drugs to be given by nominated nursing staff.

A full-time OH physician is normally in charge in both managerial and clinical areas.

Back-up clerical staff should be cost-effective for Levels 2 to 4 and are certainly necessary at Levels 3 and 4. In more complex units, where many environmental measurements and assessments are needed, an occupational hygienist may also be a member of the team.

7.7 Recruitment

For nursing staff, various methods of recruitment are available. From experience, advertising in the local newspaper normally achieves the best results. Occasionally, for example, when seeking a senior nurse to be placed in charge of the OHS, it may also be helpful to advertise in a professional journal such as 'The Nursing Times'. However, this is more expensive and in general is less fruitful.

Recruitment of general practitioners for part-time work can best be achieved by writing to the senior partner of each GP practice in the locality with a description of the job on offer. An outline job description is provided in Table 2 and should be modified to suit a particular company or site. Lists of GPs can be found in the local library.

Short-list interviews can then be arranged. If the management has no in-house medical advice available, it will, on the whole, be helpful to use the services of a medical referee. However, the prime yardstick for the local management is the usual one at such interviews, namely, 'Does the face fit?'

After selection, the successful doctor or nurse will, if taking up a full-time appointment, complete the normal company employment contract. A model appointment letter for a part-time doctor is given in Table 3.

Table 2. Job description company/unit medical adviser
(full- or part-time)

I. *Place in the Organization*:
Organization chart: (insert as appropriate)

II. *General Objectives of the Job*:
1. To advise management and, on a basis agreed with mangement, all employees on occupational health (OH).
2. To act as point of professional contact with:
   (a) local family practitioners,
   (b) local consultants and
   (c) local hospitals.
3. Whilst carrying out objectives 1 and 2, to maintain a role in the practice of clinical medicine.

III. *Fields of Responsibility*:
1. Management: To advise on:
   1.1 The level of OH service which the company should provide to meet both Government standards and the company's OH policy.
   1.2 How 1.1 should be achieved and what steps are necessary to provide this service.
   1.3 Implementation of the company's OH policy, as agreed with the company management.
   1.4 Achieving acceptance within the company of the importance of occupational medicine in the OH service.
   1.5 Health hazards within the company.
   1.6 Product safety in the company.
   1.7 Legislative changes in OH (the UK and EC).
   1.8 Scientific changes in OH.
   1.9 Health centre appointments in the company.
   1.10 Organization of OH services.
   1.11 The budget of the company OH services.
   1.12 Education:
      1.12.1 Functional training of medical staff.
      1.12.2 Management training in OH.
   1.13 Good communications with:
      1.13.1 The local medical services.
      1.13.2 All those employed in the company OH services.
2. To advise the chairman, unit heads and their employees on:
   2.1 Items 1. above.
   2.2 The existence (or otherwise) of available medical and para-medical assistance both within and outside the company.
3. To keep informed and to transmit relevant information to the chairman, management and unit heads of the results of clinical OH and research medicine.

In order to achieve this, establish and maintain liaison with:
- 3.1 The local Health and Safety Executive (e.g., EMAS and the Factory Inspectorate),
- 3.2 Local hospitals and consultants,
- 3.3 Local family practitioners and
- 3.4 The Faculty of Occupational Medicine at the Royal College of Physicians and the Society of Occupational Medicine.
4. To liaise with the nursing staff.
5. To represent the company's OH interests at external meetings, symposia and committees, where appropriate.
6. To liaise with contacts arising from the job listed under IV. below.

IV. *Contacts arising from the job*:

NB: This draft is written for a company medical adviser. For a *site* medical adviser, substitute the word 'site' (or 'unit') for 'company', as appropriate.

Table 3. Model letter of contract for part-time medical adviser

Dear Dr

It is agreed between us that from .... [date]

a) You will act as part-time Occupational Medical Adviser for us at our .... factory for ... hours a week and, in such capacity, will perform such services as and when we may reasonably require.

b) Your remuneration will be at the rate of £..... per year payable monthly on the 21st of each month, the first of which will be paid on the 21st ...... Out of your remuneration, you are to pay for the services of any doctor who acts as your locum tenens in your absence.

c) You will be entitled to five weeks' holiday each year. During your absence on holiday or incapacity due to sickness or injury, you will be entitled to your remuneration and it is agreed that, if your absence is of four weeks or more, you may be required to procure the services of another doctor or doctors to perform your services in your absence.

d) Your appointment will continue until terminated by either of us at any time giving to the other three months' notice, or your 65th (*) birthday, whichever is the earlier.

e) You will not, at any time hereinafter, except in the proper course of your duties, divulge to any third party or use any information of a secret or confidential character acquired by you during the course of your services relating to our affairs or those of any of our associated companies.

If you agree that the above correctly records our agreement, will you kindly sign the confirmation at the foot of the enclosed copy of this letter and return it to ....

Yours sincerely,

*or the appropriate company retirement age.

7.8 Medical records and reports

Medical records, which relate to the health or illness of individual employees, are strictly confidential. They may not, without the signed consent of that individual, be released, in whole or part, to any third party. The only exception to this is where release is required by a court order. In order to safeguard medical records, a secure, locked cabinet must be provided in the OH centre. All keys must remain with the OH staff.

Management can ask for any reasonable medical report on an employee and the OH staff should provide this, subject to agreement from the individual concerned. The Access to Medical Reports Act 1988 [8] requires the employee to give his or her signed consent before the report is released to a third party.

8. THE PLACE OF THE OHS IN THE ORGANIZATION

Daily functions of the OHS, such as investigation of sickness absence, return from illness and rehabilitation after injury, usually relate most closely to the personnel department. On the other hand, matters such as the design of workplaces, processes, equipment, ergonomics and ventilation lie within the scope of the technical management.

The most common arrangement places the OHS within the personnel function. The doctor may have a reporting relationship to the factory manager, personnel or technical directors whilst the nurse in charge may report directly to the site personnel manager. However, in some chemical companies, the OH and safety services report to the technical director and his factory manager. In general, it is more satisfactory for the OHS to be wholly within one function. However, the occupational physician should have access to the most senior manager on site if he should need to do so.

For the smooth running of the OHS and its integration within the company structure, it is vital that the management understand the role, functions and constraints which apply to the OH staff. As the doctor and nurse are part of the medical profession, they must abide by the ethical code and rules of their respective professional bodies. Thus, they must observe confidentiality of medical data and records. Nevertheless, they can play a full role as part of the management team without, in any way, hiding behind medical confidentiality.

The doctor and nurse must give, and be seen to give, impartial advice to both the company and the employees, particularly on issues such as sickness absence. In most cases, there are few problems where the needs of the individual and the company coincide. In a few, however, there may be a conflict of interest between the sick employee and the manager which, with diplomacy and care on the doctor's part, can be resolved.

The doctor and nurse can, by being pro-active, not only do much to prevent ill-health, accidents and absence but also become integral members of the company team, contributing to the overall business strategy and success of the undertaking. Nowhere is this more possible than in a modern chemical plant.

REFERENCES

[1] International Labour Conference 1959, Recommendation No. 112, 1959, ILO, Geneva.
[2] International Labour Conference 1985, Convention No. 161, 1985, ILO, Geneva.
[3] International Labour Conference 1985, Recommendation No. 171, 1985, ILO, Geneva.
[4] The Health and Safety at Work etc. Act 1974, (ISBN 0 10 543774 3) HMSO.
[5] The Control of Substances Hazardous to Health Regulations 1988, SI 1988 No. 1657 (ISBN 0 11 087657 1), 1988, HMSO, and subsequent amendments.
[6] The Reporting of Injuries, Diseases and Dangerous Occurrences Regulations 1985, SI 1985 No. 2023 (ISBN 0 11 058023 0), HMSO.
[7] 'Plan of Work for 1990–91 and Beyond', Health and Safety Commission, 1990, HMSO.
[8] Access to Medical Reports Act 1988, (ISBN 0 10 542888 4), 1988, HMSO.

5

Occupational health auditing

Dr M. K. B. Molyneux

SUMMARY

This chapter explains how structured-team auditing can be applied. It presents views on the principles, with particular reference to purpose, content, application in the workplace and reporting. The purpose of auditing is to compare prevailing conditions with pre-determined standards so that corrective action and improvements can be achieved. Company policy and codes of practice should define the scope and standards to be achieved. These are addressed systematically in the audit at an appropriate level depending on the type of audit, e.g. corporate, site or departmental. The findings can be quantified to make internal comparisons and to identify trends. Both the integrity of the control systems and the effectiveness of application can be monitored in this way. Reports can give sufficient detail to target specific remedial actions and more generalized information. Reference is made to commonly met variants which also have a role in occupational health practice. Lastly, views are expressed on the elements of a management system within which occupational health audits can be most effectively implemented. Seen in perspective, regular audits provide a valuable means of identifying priorities and promoting improvements towards the set standards. Adequate concentration of effort and continuity can be achieved through an audit programme linked with company business plans.

The skills required at the corporate and departmental level are multi-disciplinary, which need to be reflected in the selection of auditing teams. The collective skills need to be broad enough to appreciate the interface between health, safety and the environment and to assess the health, technological, operational and legislative aspects of the workplace. Team members need to be competent by virtue of experience and training for the task.

1. INTRODUCTION

Employers have well-defined duties to provide risk-free systems of work as far as reasonably practicable. These are stated, for example, in the Health and Safety at Work etc. Act 1974 [1] and, more specifically, in the Control of Substances Hazardous to Health Regulations 1988 [2]. Workplace factors which influence exposures are dynamic and intrinsically variable. When exposures change, so do the risks. In many cases, the effects are insidious and have long latency. For these reasons, the effectiveness of controls cannot be taken on trust; more objective means of evaluation are required.

Standards which define the requisite degree of control are to be found in modern management systems. They follow the lead set by company policy, appearing as industry codes of practice and standard operating procedures. They apply to all systems—technical, commercial, health, safety and environmental [3]. Awareness of these standards is critical to the functioning of the systems to which they apply and to auditing. Without standards, there is no objective means of assessing compliance. Thus, structured auditing is locked into modern management practice.

The experience of CIA member companies has been published in one of its major guidance booklets [4], the section on health being used to illustrate key aspects included in this chapter. Overall, the views expressed are intended to give the reader sufficient basic knowledge to set about formulating, planning and carrying out audits. Although it is not appropriate to address commercially available packages, the guidance should also assist readers to get the maximum benefit from such packages.

The examples of factory legislation quoted above stand alongside other equally important acts and statutory instruments, such as the Environmental Protection Act 1990 [5] and the Control of Industrial Major Accident Hazards Regulations 1988 [6]. While these workplace and environmental requirements differ, in detail, they still share similar principles of recognition, evaluation and control. Thus, common ground can be exploited both in the application of standards and auditing. This has led to the evolution of combined health, safety and environmental audits where the dimensions are workable. Although this chapter is concerned solely with health, the interface is recognised and is dealt with, in detail, in the CIA guidance publication [4].

2. PRINCIPLES

The principles of good practice for structured auditing are identified as follows:

— purpose,
— standards,
— content,
— application,
— reporting, and
— management.

The primary purpose of auditing, generally, is to check compliance with occupational health standards. Each audit itself also needs to have a stated purpose which reflects the selected targets.

Standards need to be set before compliance auditing can begin. The overall purpose should be embraced in policy, the detailed standards in codes of practice. Content should be derived from the standards, with similar structure and detail, and be relevant to the subject of the audit.

Application in the workplace requires competent team members, structured information gathering, judgements on compliance and recommendations for remedial action which are reasonably practicable. Reporting should be complete, streamlined and without delay, directly linking the findings with—but avoiding restating—the standard. Management systems appropriate to occupational health need to be in place and operating, i.e. policy, organization, planning and performance monitoring, including duties in law. These principles are discussed in the following sections.

3. PURPOSE

3.1 Systems and functions

A primary purpose of auditing is to assess performance against agreed standards. The key factors are:

— systems, which are necessary for implementation of the standards, and
— functions, which apply to the effective use of the systems.

The terms 'compliance' and 'performance' are inter-related, the latter placing a judgement on the degree to which compliance with the standard is achieved.

For any one audit, the purpose needs to be considered in some depth, taking account of systems and functions, direction and detail, process (or department) and population. Thus, audits can be used to check that a mechanism is in place (i.e., system) and how it is being used to meet its targets (i.e., function). They can be used to check overall compliance with policy (i.e., direction) and they can check the extent to which the policy is being implemented in the workplace (i.e., detail). They can be used to check on the application of standards to employees, contractors and the community. Where it is appropriate to carry out integrated auditing, the purpose can embrace health, safety and the environment.

The facets referred to above are not mutually exclusive, since audits can be designed to check on, for example, systems and functions, direction and detail. It is inevitable that size, complexity and organization will place constraints on the practical scope of any one audit.

3.2 Benefits

Whilst assessing compliance is clearly a primary purpose of auditing, the tangible benefits appear by way of:

— improved awareness of standards,
— recognition of shortcomings, and
— enhanced performance through improvement planning.

For example, data generated in any one audit or collection of audits can be used to indicate status and trends. Strengths and weaknesses can be identified and used as examples to improve awareness and to formulate solutions. This aspect can be exploited to greatest advantage when combined with inter-departmental and cross-sectoral audits where a wide spectrum of performance exists, can be demonstrated and explained. Where there is an improvement plan, audits can be timed and targeted to track progress and stimulate performance, such as might arise with the introduction of new legislation.

4. STANDARDS AND CONTENT

4.1 Standards
In any audit, it is first necessary to identify the standards which are to be the basis of comparison. This is the case whether the audit is to be concerned with plant and equipment or systems, procedures and operating practices.
These standards may include:

— the requirements of legislation, approved codes of practice and guidance notes,
— company policy,
— management systems and control procedures,
— operating procedures and systems of work, and
— other codes of practice and specifications.

The auditors must have a clear understanding of the relevant standards. In situations where the auditors are not familiar with the unit and its operational practices—for example, in the case of a formal audit by external auditors—it is necessary for them to gain a good awareness and understanding of local management control systems, procedures and arrangements. Normally, such information and the appropriate documentation would be given to the auditors in advance of the audit.

The established standards, systems of work and procedures should be critically reviewed by the auditors to assess their relevance to the activity and risks in question, to consider their strengths and weaknesses and to evaluate their overall quality.

4.2 Content
A prerequisite for any structured audit is the preparation of documentation which gives the topic and the standard questions against which performance can be judged. An example of a set of standard questions on agent inventories, abstracted from an audit of assessment of risk, is given in Table 1.

The full scope of occupational health auditing is indicated in the CIA guidance, which has identified 11 headings and over 70 sub-headings, covering the main aspects of the field (see Appendix 1 to this chapter).

Table 1. Examples of standard questions from a structured audit on risk assessment

SECTION 2: AGENT INVENTORIES

PART 2.1 AGENT INVENTORY

Question 2.1.1 Is the agent inventory complete?

PART 2.2 AGENTS

Question 2.2.1 Have all physical agents been listed?
Question 2.2.2 Have all biological agents been listed?
Question 2.2.3 Have all agents arising from the process/activity been included?
Question 2.2.4 Have all scrap and waste materials been included?
Question 2.2.5 Have all cleaning materials been included?
Question 2.2.6 Have all maintenance materials been included?
Question 2.2.7 Have all materials of construction of buildings, plant and equipment been included?

PART 2.3 AGENT CODES

Question 2.3.1 Are all the agent codes listed?

PART 2.4 HAZARDS OF AGENTS

Question 2.4.1 Are the hazards for each agent listed?
Question 2.4.2 Is the physical form considered for chemical substances?

PART 2.5 HAZARD DATA SHEETS

Question 2.5.1 Are all the hazard data sheets included or referenced?

PART 2.6 DATING OF HAZARD DATA SHEETS

Question 2.6.1 Are all the hazard data sheets dated?

PART 2.7 OCCUPATIONAL EXPOSURE LIMITS (OELs)

Question 2.7.1 Does the agent inventory list all published OELs?
Question 2.7.2 Is the source of each OEL clearly identified?
Question 2.7.3 Are internal working limits for appropriate agents listed?

PART 2.8 JOB TYPE AGENT INVENTORY

Question 2.8.1 Are job type agent inventories clearly listed?

The following eight of the eleven headings have a direct bearing on the risk of disease or other harmful effects:

- organization and management,
- hazards and risks to health,
- control of exposure,
- monitoring of exposure,
- health surveillance,
- emergency response,
- information, instruction and training, and
- records.

The two headings below relate to primary health care:

- first aid, and
- health promotion.

The final heading is:

- internal audit,

emphasizing its place within the totality of occupational health practice.

Collectively, they represent the breadth of the multi-disciplinary activities as presently seen by industrial practitioners.

4.3 Subject details

The 70 sub-headings, listed in full in Appendix 1, give an insight into the subject details. Examples of the aspects covered are:

- methods of risk assessment and the assessment programme (Heading 2), safe operating procedures and occupational exposure limits (Heading 3), exposure data (Heading 4);
- control and monitoring of exposure and application of reasonable practicability (Headings 3 and 4);
- health surveillance, systems and application, fitness for work and health education (Headings 5 and 7);
- plans, communication and procedures for major industrial hazard emergencies (Heading 8);
- programme and system for auditing (Heading 11).

The penultimate example serves as a reminder that there are some topics, such as major hazards, where there may be a joint health, safety and environmental involvement which has to be recognized in all three areas of practice.

The last example illustrates the need to audit auditing as one of the control procedures.

4.4 Business activity

Industrial organizations are generally divided into departments and stratified to a degree, depending on size and complexity. The extremes are the single site company and the multi-national. However, whatever the size or complexity, from the outset, the auditing process needs to be targeted at a stated activity and organizational level. Activity is typically defined by product or output while level is determined by line management structure, ranging from central office functions to, say, production units.

The extent to which an audit can deal with the detailed effectiveness of an organization is directly influenced by the level. An audit of corporate activities, for example, could not be expected to yield detailed information on the effectiveness of controls in workplaces at remote manufacturing locations. It would, however, be expected to identify and place a judgement on awareness, lines of communication, management controls, policy and business plans. A detailed view of engineering controls in the workplace can be obtained by visiting a production unit and thoroughly scrutinizing the built-in controls, its procedures and the competence of its employees to use them.

The two examples quoted above have one thing in common; they both require a fully-documented, structured audit requiring dedicated site visits. There are other situations where minimal documentation may be required, where the audits are self-administered and yet serve important purposes.

5. PRACTICAL APPLICATION

From the above, it will be seen that properly prepared audits will have a purpose, title, clearly defined headings and detailed questions, all attributable to documented standards. In order to apply audits, it is then necessary to collect relevant information, form judgements on compliance and remedial action, record the findings and report results.

5.1 Audit team

In any audit plan, a portion of the auditing will be 'self-auditing', i.e., carried out by the local management, at whatever level, and their staff. Depending on the scope of the proposed audit, the team can be supported by specialists in safety, occupational health, environmental protection and other disciplines, to provide specific expertise or a degree of independence. Occasionally, it may be appropriate to have an 'external' audit carried out by auditors who are independent of the unit, of the site or even the company to be audited.

Auditors, at all levels, must fully understand their tasks and be competent to carry them out. They must have the status, experience and knowledge of standards and systems appropriate to the audit concerned. In addition, they must have the skills and judgement to observe and evaluate performance against relevant standards, identify deficiencies and prescribe remedial action. Training should be given, as necessary, to ensure that auditors have the appropriate competence for the task.

5.2 Information collection

Auditors should seek objective evidence on the degree of adherence to, and compliance with, relevant standards and procedures in operating practice. Evidence for compliance may be gathered in the following ways:

— personal observation,
— inspection of records,
— discussion with local management, supervisors and operators,

- audit questionnaires,
- testing of systems and procedures,
- inspection of plant and equipment
- local environmental effects, and
- health effects.

It is often helpful to use a checklist or questionnaire as an aide-memoire to ensure that all aspects are covered and to provide a structure to the audit. The 70 sub-headings listed in Appendix 1 can be used to serve this purpose.

5.3 Performance rating

Information should be critically reviewed, taking into account its relevance to the activities and risks in question, to consider strengths and weaknesses and evaluate overall quality. Many judgements will be subjective and liable to individual bias. Differences of opinion need to be reconciled so that the recipient can be presented with a realistic, detached view. The knowledge, skills and experience of the audit team are crucial in reaching this end-point.

The least structured method of performance rating is by intuition where, given the topic, the auditors comment on performance in their own personalized manner. In audits of this type, the auditors may also formulate the questions which may or may not be documented (see the 'Team of One' below).

The most structured method is to compare observations with a performance scale and report the point on the scale. This has the advantage of being standardized and, for anything less than 100% compliance, also indicates the remaining targets to be achieved. This scheme is adopted in the CIA auditing guidance [4] where each of the 70 occupational health topics is classified in a four-point performance scale. An example of the scale for the topic of Control of Exposure is given in Appendix 2 to this chapter.

Between these extremes are variants which tend towards one or other. The intuitive method, for example, can be structured with a fully documented audit where performance is scored over a simple scale between 0 and 5, 10 or 100. Such scores can be weighted to reflect views on relative importance, which is a feature of the more complex evaluation and reporting systems.

5.4 Records and reporting

A standard method for recording observations, judgements on compliance and recommendations, during the course of the practical audit, greatly assists the preparation of the report.

If the standard is well documented and available to all parties (as is recommended), the reporting process can be automated, thereby minimizing reporting time and effort and ensuring a standard approach. In this way, the narrative can be kept to a summary of the main findings plus recommendations for remedial action.

Otherwise, the auditors have the task of repeatedly justifying every judgement on compliance with an explanation of the standard, which is to be avoided.

The recommendations should form the input to the improvement plan, as described in Section 7 'Management' (below).

Structured auditing lends itself to automated reporting where software can be used to generate and format the questions, process the scores and record a narrative on findings and remedial action.

5.5 Results and analysis of an audit report

The information presented in this section illustrates data which might be obtained from an audit of the assessment of risk to health such as would be required for implementation of the COSHH Regulations 1988.

The audit was based on 109 standard questions in 42 parts in 6 sections (see below) which were prepared prior to the audit. The six sections were:

— administration,
— agent inventories,
— job and task inventories,
— appraisal of procedures,
— appraisal of facilities, and
— appraisal of tasks.

Eleven locations were audited and performance was ranked on a scale 0–10 where:

0 = no action yet taken,
3 = action being taken to implement,
6 = implementation substantially complete and
10 = implementation fully satisfactory.

Results, as scores (%), are presented as follows:

Table 2—for each location,
Table 3—for each section of the audit,
Table 4—for each location for each section of the audit,
Table 5—for each location for each part of the audit,
Table 6—for each part of the audit, overall.

From these results the performance of these locations could be assessed as follows:

— The best location had an overall score of 99%, the worst 82%; most of the remainder were above 91%.
— The best scores were obtained for appraisals of procedures and facilities and administration; the worst were for agent inventories and task appraisals.
— Location D had the most robust profile, followed by location B. Location A had the weakest profile across Sections 1–6, followed by location C.
— Topics covered by Parts 2, 3 and 6 exhibited the most common cross-location weaknesses.
— Of the 42 parts, 17 obtained the maximum possible score. The lowest scores were obtained in hazard (identification), job type coding and remedial action (documentation), all of which were below 70%.

Overall, this might be regarded as a favourable outcome with some clear pointers to areas where improvements would be required.

Table 2. Scores (%) for each location

Location	Overall Score
A	86.2
B	98.0
C	81.7
D	99.0
E	94.2
F	91.3
G	91.3
H	96.0
I	97.2
J	94.8
K	91.5

Table 3. Scores (%) for each section of the audit

SECTION 1	ADMINISTRATION	95.5
SECTION 2	AGENT INVENTORIES	86.0
SECTION 3	JOB and TASK INVENTORIES	90.8
SECTION 4	APPRAISAL OF PROCEDURES	98.9
SECTION 5	APPRAISAL OF FACILITIES	96.6
SECTION 6	APPRAISAL OF TASKS	89.2

Table 4. Scores (%) for each location for each section of the audit

	A	B	C	D	E	F	G	H	I	J	K
SECTION 1	100	100	50	100	100	100	100	100	100	100	100
SECTION 2	65	94	65	100	100	82	82	94	94	88	82
SECTION 3	83	100	83	100	100	83	83	100	100	100	67
SECTION 4	96	100	92	100	100	100	100	100	100	100	100
SECTION 5	90	100	100	100	93	100	100	93	100	87	100
SECTION 6	83	94	100	94	72	83	83	89	89	94	100

Sec. 5] **Practical application** 81

Table 5. Scores (%) for each location for each part of the audit

	A	B	C	D	E	F	G	H	I	J	K
PART 1.1	100	100	0	100	100	100	100	100	100	100	100
PART 1.2	100	100	100	100	100	100	100	100	100	100	100
PART 1.3	100	100	100	100	100	100	100	100	100	100	100
PART 1.4	100	100	0	100	100	100	100	100	100	100	100
PART 2.1	100	100	0	100	100	100	100	100	100	0	100
PART 2.2	71	86	43	100	100	86	86	100	86	100	100
PART 2.3	0	100	100	100	100	0	0	100	100	100	0
PART 2.4	100	100	50	100	100	100	100	50	100	100	100
PART 2.5	100	100	100	100	100	100	100	100	100	100	100
PART 2.6	100	100	100	100	100	100	100	100	100	100	100
PART 2.7	33	100	100	100	100	67	67	100	100	67	67
PART 2.8	0	100	100	100	100	100	100	100	100	100	0
PART 3.1	100	100	100	100	100	100	100	100	100	100	0
PART 3.2	0	100	100	100	100	0	0	100	100	100	0
PART 3.3	100	100	0	100	100	100	100	100	100	100	100
PART 3.4	100	100	100	100	100	100	100	100	100	100	100
PART 3.5	100	100	100	100	100	100	100	100	100	100	100
PART 3.6	100	100	100	100	100	100	100	100	100	100	100
PART 4.1	67	100	100	100	100	100	100	100	100	100	100
PART 4.2	100	100	100	100	100	100	100	100	100	100	100
PART 4.3	100	100	100	100	100	100	100	100	100	100	100
PART 4.4	100	100	100	100	100	100	100	100	100	100	100
PART 4.5	100	100	100	100	100	100	100	100	100	100	100
PART 4.6	100	100	100	100	100	100	100	100	100	100	100
PART 4.7	100	100	67	100	100	100	100	100	100	100	100
PART 4.8	100	100	100	100	100	100	100	100	100	100	100
PART 4.9	100	100	0	100	100	100	100	100	100	100	100
PART 5.1	75	100	100	100	100	100	100	100	100	100	100
PART 5.2	100	100	100	100	100	100	100	100	100	75	100
PART 5.3	100	100	100	100	100	100	100	100	100	100	100
PART 5.4	80	100	100	100	100	100	100	80	100	100	100
PART 5.5	100	100	100	100	100	100	100	67	100	100	100
PART 5.6	100	100	100	100	100	100	100	100	100	100	100
PART 5.7	75	100	100	100	100	100	100	100	100	25	100
PART 5.8	100	100	100	100	60	100	100	100	100	100	100
PART 6.1	100	0	100	100	100	100	100	100	100	100	100
PART 6.2	50	100	100	100	100	100	100	100	100	100	100
PART 6.3	100	100	100	100	100	100	100	100	100	100	100
PART 6.4	100	100	100	100	100	100	100	100	100	100	100
PART 6.5	100	100	100	100	50	100	100	100	100	100	100
PART 6.6	100	100	100	100	80	80	80	80	100	100	100
PART 6.7	33	100	100	67	33	33	33	67	33	67	100

Table 6. Scores (%) for each part of the audit

PART 1.1	Identification of Assessment Unit	97.9
PART 1.2	Organization of Assessment Unit	100.0
PART 1.3	Plan of Assessment unit	100.0
PART 1.4	Description of Process Activity	90.9
PART 2.1	Agent Inventory	81.8
PART 2.2	Agents	87.0
PART 2.3	Agent Codes	64.0
PART 2.4	Hazards of Agents	90.9
PART 2.5	Hazard Data Sheets	100.0
PART 2.6	Dating of Hazard Data Sheets	100.0
PART 2.7	Occupational Exposure Limits	81.9
PART 2.8	Job Type Agent Inventories	81.9
PART 3.1	Job Type Inventories	90.9
PART 3.2	Job Type Codes	63.6
PART 3.3	Numbers	90.9
PART 3.4	Job Descriptions	100.0
PART 3.5	Task Lists	100.0
PART 3.6	Shifts and Hours	100.0
PART 4.1	Work Method and Emergency Action	97.0
PART 4.2	Information	100.0
PART 4.3	Instruction and Training	100.0
PART 4.4	Maintenance and Housekeeping	100.0
PART 4.5	Personal Hygiene	100.0
PART 4.6	Waste Disposal	100.0
PART 4.7	Exposure Monitoring and Health Surveillance	100.0
PART 4.8	Supervision	100.0
PART 4.9	Process Change	90.9
PART 5.1	Design and Layout	97.7
PART 5.2	Construction	97.7
PART 5.3	Containment	100.0
PART 5.4	Local Exhaust and General Ventilation	96.3
PART 5.5	Vents and Exhausts	97.0
PART 5.6	Drains	100.0
PART 5.7	Thermal and Visual Environment	90.9
PART 5.8	Amenities	96.3
PART 6.1	Description of Tasks	90.9
PART 6.2	Identification of Agents	95.4
PART 6.3	Routes of Exposure	100.0
PART 6.4	Frequency and Duration of Exposure	100.0
PART 6.5	Estimates of Exposure	95.4
PART 6.6	Control Measures	92.7
PART 6.7	Remedial Action	60.5

6. OTHER TYPES OF AUDIT

The principles and practices so far described apply to structured audits which are typically carried out by a team on sites or operations of some size. The results are detailed and the audits carried out regularly but relatively infrequently.

There are other requirements which are met by audit methodology where the scope is small but immediate feedback is needed or where a more superficial assessment is adequate. Examples are:

— regular inspection of engineering or procedural controls by operating staff, and
— annual situation reports on compliance with industry policy and practice.

6.1 Regular inspections

Regular inspection normally takes the form of a routine tour of workplaces by trained and experienced staff to check that systems are operating and that safe methods of work are being followed. Relevant features are normally listed and methodically checked. Any deviations or failures are noted in the departmental log and reported for remedial action. The standard is defined by the standard operating procedures.

Weekly audits of this type are specified in the Approved Code of Practice on the Control of Substances Hazardous to Health Regulations 1988 [7] (Paragraph 50), applying to all controls in the workplace, and in the HSE guidance booklet on local exhaust ventilation [8] (Paragraph 30).

6.2 Situation reports

The annual situation report is typically derived from a questionnaire which is self-administered by the manager and collated centrally. The outcome is a broadbrush view of performance compared with a standard with which the manager is familiar. The quality of the response depends largely on the flow of information from departmental managers to the site manager, through local audit programmes. While this type of audit can be subject to the greatest error, it is a practicable means of obtaining a corporate or industry view.

This method is typically applied to occupational health performance indicators, an example of which is shown in Appendix 3. This reflects current thinking within CIA where performance is judged for each of the 11 subject headings (given in Appendix 1), the one exemplified being for No. 3, 'Control of Exposure'.

6.3 The team of one

Of the many possible audit variants, one other warrants comment. This relates to the role of the skilled and experienced individual auditor who may independently carry out audits in defined areas. This is likely to be most appropriate for small companies who need a periodic, independent audit coupled with guidance on the introduction of an internal audit programme. In such cases, the standard is most likely to be set initially by the auditor, based on personal knowledge and experience. Nevertheless, it needs to be clearly stated and documented, possibly leading to development of the company's own standard.

7. MANAGEMENT

Auditing is part of all quality systems which include:

— policy,
— organization,
— improvement planning, and
— evaluation of performance

which apply as much to occupational health as to any other commercial activities. Auditing is part of the evaluation of performance, showing where improvements can be made, helping to identify causes and specify action which is reasonably practicable.

While the principles and practice are generally agreed, audits are inherently fault-seeking, they can attract punitive action and are not instinctively welcomed by recipients. This background can undermine the execution and outcome of an audit programme. However, these and other problems can be avoided by careful preparation and management.

Although essential points have already been covered under 'Principles' (Section 2 above), the following are worth emphasizing:

— Standards should be known by all parties and the questions on which the audit is based should be circulated before the audit.
— Locations being audited should be actively engaged in the planning, timing, documentation and execution of the audits.
— The scope, duration, timing and frequency should aim to derive maximum benefit with minimum inconvenience.
— All answers, good and bad, should be recorded using a standard format and be agreed on site.
— The corporate and local management systems should have a clearly stated policy and organization, plus procedures for improvement planning and implementing health objectives.
— Standards should be sufficiently well defined and known, to allow for streamlined and automated reporting.
— The recipient of the report should have the power and means to implement the recommendations according to priority.
— The period between auditing and reporting should be kept to a minimum.
— A balance between internal and external (third party) audits needs to be maintained to ensure compatibility with site, company and external standards.

There are clear implications in the above list on what to avoid and what to aim for. Thus, there is a strong case for avoiding questions which are unrepresentative of the standards, prolonged and laborious audits, unscheduled audits, ambiguous recording of answers and delayed reporting.

Attention should be focused on the ultimate benefit of auditing which is improved performance, culminating in a reduction in waste, time, effort and cost. The process is intended to be informative, thus encouraging participation.

While each audit has something positive to contribute, comparisons between consecutive audits can indicate trends while comparisons between locations can be used to detect differences in performance. Both are essential inputs to an overall strategy of improvement.

8. CONCLUSIONS

Occupational health auditing provides a means of checking compliance with legal and company standards. It can be applied to systems and functions, to employees and others at risk. Among the benefits is enhanced performance which is an integral part of quality management. Audit standards can be set which give auditors a clear understanding of objectives. The structure and detailed content should be documented as a prerequisite, relating to organization, hazards and risks, control of exposure and other critical areas. From the outset, the audit needs to be targeted at a stated activity and level, giving due regard to corporate, local and technical aspects of the workplace, as appropriate. Audit plans should recognize the requirement for local self-audit, corporate audit and external audit and ensure that auditors are competent for the task. Objective evidence should be sought by personal observation using a checklist where appropriate. Observations can be ranked or scored to identify clearly strengths and weaknesses and to assist reporting. Maximum benefit can be gained by applying auditing principles to regular inspections and situation reports. Whereas a team approach is normally required, due recognition should be given to the role of the independent auditor, particularly for small companies. The whole process needs to be managed as part of company policy, organization and improvement planning, emphasizing the benefits in reduction of waste, time, effort and cost.

REFERENCES

[1] 'Health and Safety at Work etc. Act 1974' (ISBN 0 10 543774 3) 1974, HMSO.
[2] 'Control of Substances Hazardous to Health Regulations 1988', SI 1988, No. 1657, 1988, HMSO and subsequent amendments.
[3] 'Monitoring Safety', HSE Occasional Paper Series OP9, 1985, HMSO.
[4] 'Guidance on Safety, Occupational Health and Environmental Protection Auditing', (ISBN 0 900623 551) Chemical Industries Association, 1991.
[5] 'Environmental Protection Act 1990', HMSO.
[6] 'Control of Industrial Major Accident Hazards Regulations 1988', SI 1988, No. 1462, 1988, HMSO.
[7] 'Approved Code of Practice on the Control of Substances Hazardous to Health Regulations 1988', Third Edition, L 5, 1991, HMSO.
[8] 'The Maintenance, Examination and Testing of Local Exhaust Ventilation', HSE Guidance Booklet, HS(G) 54, 1990, HMSO.

APPENDIX 1: HEADINGS AND SUB-HEADINGS FOR OCCUPATIONAL HEALTH AUDITING [Reproduced from the CIA Guidance publication [4]].

HEADINGS

1. Organization and management
2. Hazards and risks to health
3. Control of exposure
4. Monitoring of exposure
5. Health surveillance
6. First aid and initial treatment
7. Health promotion and prevention of illness
8. Emergency response
9. Information, instruction and training
10. Records
11. Internal audit

SUB-HEADINGS

1. Organization and management

Policy	Statement of objectives, priorities, programmes, procedures and delegation to individuals by name and/or position.
Standards	Legal and ethical codes of practice. Monitoring of performance and progress. Review of changing needs.
Management structure	Existence of formal health and safety committee with named site focal point. Link with specialists and demonstration of leadership.
Communication	The nature of lines of communication with employees and contractors.
Community	Geographical location of plant and proximity to inhabitants.
Procedures	Written procedures for emergency, waste, emissions, control of workplace hazards and risks plus product stewardship.
Quality assurance	Systems for external and internal quality control e.g. for measurements.
Resources and competence	Training schedules linked with qualifications and facilities.
Project review	Procedures for assessing new projects, acquisitions and disinvestments.
External contact	Participation with professional bodies, industrial and trade associations and government departments.

2. Hazards and risks to health

Process	Documentation of jobs and tasks. The complexity of the activities and their scope, including ergonomic aspects such as lifting.
List of agents	Record of chemical, physical and biological agents plus ergonomic hazards including manual handling, repetitive motion and visually demanding tasks. A system for updating.
Hazard information	Access to sources of information and their quality. Competence to evaluate data, the format of data sheets and dissemination.
Surveys of exposure	Ability to carry out investigations, to collate and interpret data. The competence and availability of resources with access to methods subject to quality assurance.
Risk assessments	Documented procedure linked to a programme giving priorities. System for remedial action and for ensuring progress. Existence of review procedures, linked to health surveillance and project review.
Local history	History of harmful effects on the site including notifiable diseases, reported incidents and results of any epidemiology.
Records	See section 10.

3. Control of exposure

Occupational exposure limits (OELs)	Access to up to date published and in-house exposure limit values including biological exposure limits (BELs) and standards for manual handling.
Strategy	A systematic approach stating the principles, priorities, the programme and provision for improvement planning.
Compliance	Evidence of compliance with relevant OELs and other standards.
Plant, process	Design, construction, engineering and ergonomic standards, including workstation design and use of hand tools.
Fixed controls	Degree of containment, provision of ventilation. Scope of labelling, signs and signals.
Maintenance	The maintenance, examination and testing of control measures as defined by company and legal requirements.
Amenities	Provision of amenities for washing, changing, storage of work clothing and personal protective equipment (PPE).

Procedural controls	Written standard operating procedures, including PPE programmes and personal hygiene requirements. Permits to work for potential high risk activities.
Decontamination	Availability of equipment, plant and competence of employees.
Supervision	Awareness of supervision and effectiveness of leadership.
Records	See section 10.

4. Monitoring of exposure

Programmes	In line with statutory requirements, good practice, company policy and as indicated by risk assessments, including setting of priorities.
Protocols	Written procedures covering relevance, feasibility, content and strategy of field and analytical work.
Methods	Existence of a quality assurance scheme and method validation.
Criteria	Selection of, and access to OELs, BELs, guidance notes and other standards.
Results	Observation of trends linked with health surveillance. Procedure for feedback of information, reporting and access by employees.
Remedial action	Written procedure for defining and identifying abnormal results and taking action.
Records	See section 10.

5. Health surveillance

Criteria	Requirements as defined by statute, acceptable practice or by company policy together with rationale as indicated by risk assessments.
Protocols	A written record of the content of surveillance procedures. Communication and consultation with the workforce.
Scope	Application of health surveillance to employees, contractors and others involved.
Methods	Procedures used, i.e. for review of records. Monitoring for absence attributed to ill-health and injury. Use of health questionnaires, inspection by responsible persons, examinations by health practitioners and clinical investigations.

Results	Procedure for communication with management, the workforce collectively and individuals. Recognition of the need for and maintenance of confidentiality. Existence of a link to workplace monitoring. Conformity with a statutory reporting requirements.
Records	See section 10.

6. First aid and initial treatment

Arrangements	Requirement as defined by legislation and policy, number of first aiders and suitability of general and specific training. Existence of standing orders coupled with hazard data, treatment cards and links with health and safety functions.
Injuries	Facilities for treatment, referral and follow-up.
Ill-health	Facilities for treatment, referral and follow-up.
Equipment	Inventory of dressings and instruments, if any. Definition of responsibility by name and procedures for maintenance.
Drugs	Existence of a practice inventory. Security procedures and control of drugs. Written standing orders for nursing and paramedical staff.
External agencies	Arrangements with local hospitals, visiting specialists, general practitioners and social services. Access to and suitability of information resources.
Rehabilitation	Statement of policy. Communication with line management, personnel function and workforce.
Records	See section 10.

7. Health promotion and prevention of illness

Health education	Written programme giving system of communication and feedback of information as appropriate. Availability and suitability of resources.
Health screening	Compliance with company policy.
Foreign travel	Protocol for screening, advice, prophylaxis and immunization.
Counselling	Provision of services to individuals and groups. Observance of confidentiality. Suitability of resources.

Catering and general environment	Standard of catering facilities, food handling, changing rooms, showers, toilets and general housekeeping. Compliance with company and legal requirements.

8. Emergency response

Plan	Incorporation of health services into site plan with provision for review. Responsibilities by title and liaison with health emergency services. Level of employee awareness, systems of communication.
Equipment	Hardware and its maintenance.
Training	Programme, content and frequency of retraining. Liaison with external resources.
Rehearsals	Frequency of rehearsals and level of performance. Involvement of emergency services, procedures and review.

9. Information, instruction and training

Content	Available sources of information and their quality.
Scope	Programme of information, instruction and training in line with company and legal requirements. Induction courses for employees, contractors and specialists. Retraining and refresher courses.
Methods	Provision for internal and external training as appropriate.
Standard operating procedures (SOPs)	Content and scheme for review.
Awareness	Degree of awareness at all levels including knowledge of relevant developments.
Records	See section 10.

10. Records

Statutory	Minimum requirements as defined, e.g. by COSHH and RIDDOR* including access to medical reports.

* COSHH: Control of Substances Hazardous to Health Regulations 1988, RIDDOR: Reporting of Injuries, Diseases and Dangerous Occurrences Regulations 1986.

System	Scope and completeness of agent inventories (including quantities) and hazard data. Availability of records on hazard evaluation, processes, jobs, tasks, worker tracking, risk assessments, maintenance, examination and testing of controls including PPE, monitoring, health surveillance, information, instruction, training, audits and contractors.
Format	Lay-out, intelligibility and retrievability.
Compatibility	Links between different records within the system and between systems where appropriate.
Storage	Procedure for storage and archiving, naming the custodian and systems for maintaining security, control and confidentiality.
Access	Facilities for employees to see records in accordance with legal obligation and company practice.
Retrieval	Locating and retrieving archived records. Ease of analysis, reviewing and reporting (dissemination) of findings.
Review	Procedure for identifying trends in exposure. Linkage with new information, harmful effects and reported incidents plus facilities to incorporate into epidemiological studies.

11. Internal audit

Compliance	Compliance with Health and Safety at Work Act, other relevant legislation and company requirements.
System	Documentation of system and its suitability.
Programme	Definition of scope, frequency and procedure for audit. Availability of periodic status reports.
Reporting	Documentation of results and recommendations. Identification of trends from repeated auditing. Reporting to management.
Action	Procedure for taking action and effectiveness of follow-up.
Records	See section 10.

APPENDIX 2: PERFORMANCE RATING FOR CONTROL OF EXPOSURE (Sub-Heading 3 of Appendix 1)

3. Control of exposure

	1	2	3	4
Occupational exposure limits (OELs)	Little or no knowledge of OELs.	General awareness of published values. No 'in-house' OELs.	OELs available for all inventory materials. Where published values not available 'in-house' OELs set. OELs explained in hygiene training programme.	'In-house' OELs fully documented. OELs obtained before introduction of new materials on to plant.
Strategy	No written strategy.	A general understanding of exposure control principles and priorities but no written strategy on site improvement.	Written strategy assigning priorities and responsibilities for monitoring programmes and site improvement plan.	Strategy includes contractors and toll (contract) manufacturers. Review of strategy at least annually.
Compliance	No monitoring or assessments of exposure.	Monitoring of high hazard materials. Some assessments of exposure	Programme identifies monitoring priorities. Action levels assigned. Evidence of satisfactory completion of assessments and actions.	Management kept regularly informed of compliance status. Where appropriate, statistical summary of data presented.
Plant, process	Risks to health not assessed at plant design stage.	Informal involvement of health personnel at design stage. HAZOPs[a] for high risk processes.	Formal involvement of health personnel at design stage. HAZOPs up to date for all new and high risk processes. All plant design adequately documented and up to date.	Company standards defined and documented.
Fixed controls	Little use of fixed controls to control exposure. Heavy reliance on PPE.[b] Plant equipment and controls not identified.	Some control of exposure by containment and/or local exhaust ventilation. Critical plant and equipment controls clearly identified.	Control of exposures by engineering solution, where practicable. Plant and equipment controls clearly identified by appropriate use of labels, signs and signals.	Periodic review of plant 'fitness for purpose' taking into account hygiene issues.
Maintenance	No formal attention given to maintenance, examination and testing of control measures. No established procedures or awareness of company or statutory requirements.	Awareness of statutory requirements. Maintenance programmes and documentation partially complete.	Formal programmes cover all statutory requirements relating to maintenance of control measures.	Systems set up for management of a preventative programme.

3. Control of exposure (continued)

	1	2	3	4
Amenities	Few or no amenities provided.	Some facilities for changing and washing. Inadequate cleaning of amenities and work clothing.	Adequate facilities provided for washing, changing and storage of work clothing and PPE. Provision of work clothing, laundering and cleaning adequately addressed.	Inspection and audit to ensure adherence to procedures.
Procedural controls	Some written procedures available.	SOPs[c] available but key OH[a] issues omitted. Permit to work systems established but not consistently used. Informal controls on issue and use of PPE.	SOPs cover all relevant OH issues. Formal PPE programme addresses selection, use, storage, maintenance, issue and records. Permit to work system established, understood and used.	All procedures subject to regular review, update and audit.
Decontamination	No awareness of need for decontamination procedures.	Some procedures established but not always adequate, well understood or consistently used.	Adequate procedures and equipment provided for decontamination of equipment and personnel. Procedures well understood and followed.	Periodic evaluation ensuring efficiency of decontamination procedures.
Supervision	Supervisory responsibilities not assigned.	Insufficient supervisory staff assigned. No specific training on occupational hygiene issues.	Adequate provision of supervisory staff with formal training in OH issues. Hygiene problems addressed in regular production meetings.	Supervisors receive specialised training on control of risks

[a] HAZOPs = hazard and operability studies.
[b] PPE = personal protective equipment.
[c] SOPs = standard operating procedures.
[d] OH = occupational health.

APPENDIX 3: CIA RESPONSIBLE CARE INDICATOR OF PERFORMANCE IN OCCUPATIONAL HEALTH

Guidance notes—data on occupational health information

This performance indicator is based on the occupational health (OH) section of CIA publication 'Guidance on safety, occupational health and environmental protection auditing' published in March 1991 (ISBN 0 900623 55 1) which is available from the CIA Publications Office.

For each of the following 11 headings:
PLEASE TICK ONE BOX ONLY
that best describes the occupational health standard attained by your site.

3. Control of Exposure

Notes: How would you best describe your compliance with the control of exposure on your site:

- A Recommendations on control from assessments not implemented and no data on compliance with Occupational Exposure Limits (OELs).
- B Recommendations partly implemented but dependent on Personal Protective Equipment (PPE). Some evidence of compliance with OELs.
- C Recommendations largely implemented as far as is reasonably practicable but with some dependence on PPE. Substantial evidence with compliance with OELs.
- D Recommendations fully implemented as far as is reasonably practicable with minimal dependence on PPE and with complete evidence of compliance with OELs.

6

Hazard assessment

Dr F. M. B. Carpanini

SUMMARY

Assessment of the nature and severity of toxic hazards is fundamental to responsible risk management of chemicals which may come into contact with people. In the workplace, such assessment forms the basis of the development of appropriate controls and measures to protect the health of the workforce.

In this chapter, the process of chemical hazard assessment is described and some of the underlying principles and practices explained. The need to ensure the participation of appropriate specialists, such as toxicologists, physicians and occupational hygienists, is emphasized in view of the complexity of the assessment process and the frequent limitations in the data which are available.

1. INTRODUCTION

It is theoretically possible for all chemicals to be manufactured and used in industry without causing harm to the workforce or the community at large. Simplistically, one way of achieving this would be to ensure that the substances are completely isolated from contact with people and the environment. Even if it were possible to prevent accidental failures in containment, such an option is impractical since, ultimately, in some form or other, chemicals still have to come into contact with people and their surroundings in the form of the desired end-products. Furthermore, the high costs of providing a high level of containment for all manufacturing processes would render uneconomical a large proportion of the many benefits which accrue to society from chemical products.

A more practical way of achieving the safe use of chemicals is through recognition of their potential to affect adversely the health of individuals with whom they come into contact and to employ appropriate and effective control measures so as to limit that contact—or exposure—to levels which are safe.

The study of the adverse effects of substances on living organisms is known as toxicology. For many people, toxicology evokes thoughts of poisons and, in particular, substances which are lethal. Certainly, in its early days, devotees of toxicology included the infamous Catherine de Medici who is alleged to have been an accomplished poisoner. Nowadays, the discipline of toxicology makes use of a broad spectrum of skills from the natural, biological and medical sciences and is concerned with predicting *all* adverse effects on people, not only those which are fatal.

A fundamental principle of toxicology is that all substances are toxic; that is to say that every chemical has the potential to cause harm. This is true even of water and oxygen, both of which are essential to life but can, under certain conditions of exposure, cause severe harm and even death.

Another principle, true for most (if not all) substances, is that, in spite of their intrinsic potential to cause harm, the nature and severity of adverse effects which they can produce depend on the magnitude, route and duration of exposure. It should, therefore, be possible to define a level and form of exposure below which toxic effects will not occur. This subject is covered in detail in Chapter 10.

Experience has shown that not all chemicals have the same potential to cause harm. Some, like hydrogen cyanide or arsenic, are recognised as being highly poisonous, whilst others, like calcium carbonate or, indeed, water, are relatively innocuous. It follows that, in order to assess the hazards of chemicals in relation to their use, it is necessary to consider not only the toxicity of those substances but also the degrees and types of exposures which can occur. Knowledge of the workplace conditions facilitates the development and adoption of appropriate measures of control for minimizing the risks of the workforce being affected by the toxic hazards. These are the principles which underpin the Control of Substances Hazardous to Health (COSHH) Regulations 1988 in the UK, the purpose of which is to protect the health of workers by requiring hazardous substances in the workplace to be identified and the potential risks to health associated with the use of those substances assessed.

Comprehensive guidance on the COSHH Regulations and their implementation is given in the associated general Approved Code of Practice (ACoP) and a series of guidance booklets on specific aspects of COSHH published by the Chemical Industries Association. Of particular relevance to hazard assessment is the CIA booklet on the 'Collection and Evaluation of Hazard Information (Regulation 2)'. (See also Appendix II at the end of the book.)

Hazard assessment is discussed below in terms of the following stages:
(a) definition of substances concerned,
(b) consideration of exposure characteristics,
(c) review of toxicity data,
(d) evaluation of hazards, and
(e) periodic review.

2. DEFINITION OF SUBSTANCES CONCERNED

An essential first stage is to define, as precisely as possible, the identity, composition (including impurities), physical and chemical properties and form of the substance(s) to which individuals may be exposed in the workplace. Substances cover a wide range of materials, pure and impure, simple or complex mixtures, single chemical compounds, natural or synthetic products and may be in the form of solids, liquids, gases, dusts, fibres, fumes or aerosols.

The importance of this first stage cannot be over-emphasized since these factors can dramatically alter the conclusions that may be reached in the final assessment of hazards and risks. Some substances can be relatively risk-free in solid or liquid form but may present a high risk to health in the form of dusts, aerosols or vapours, e.g. welding rods compared with welding fume. Similarly, silica may be relatively harmless in its amorphous form compared with its toxicity when in crystalline or fibrous forms. There are occasions when it is difficult, if not impossible, to define, the substances in precise chemical terms, especially those which are produced to conform with technological specifications as opposed to chemical criteria, e.g. viscosity or boiling point. In certain cases, it may be appropriate to define the materials in terms of the process conditions and source materials from which they have been derived.

In considering the range of chemicals to which the workforce may be exposed, Table 1 may be useful. The range of substances comprises not only materials brought

Table 1. Range of sources of chemicals to be considered

Input (suppliers)	Workplace	Output (customers/ environment)
Raw materials	Intermediates	
Additives	Products	Products
Catalysts	Formulations	Formulations
Components	Impurities	(Impurities)
Research (trial) materials	Off-specification materials	Wastes
Maintenance materials	Residues	
	Spent catalysts	
	By-products	

onto the site, which are relatively easy to identify, but also those which may be produced in one process or another (even maintenance) and to which exposure can occur.

3. EXPOSURE CHARACTERISTICS

Having established, as accurately as possible, the identity of the substances to be evaluated, the next stage is to define the characteristics of the exposures which are likely or possible. Key elements to be defined include the magnitude of exposure, the route of absorption into the body, the frequency and duration of each exposure as well as the total period over which the exposure is likely to occur. For industrial chemicals, the route of absorption will mainly be via inhalation or skin (eye) contact but may include oral (by food contamination) or injection (hydraulics).

4. REVIEW OF TOXICITY DATA

In the third stage of the hazard assessment process, the available information on the toxicological properties of the substance is collected and critically reviewed. The toxicological data available for chemicals are extremely variable in quality and quantity. Human data should, in theory, provide the best insight into the likely effects in people, but often such data are derived from anecdotal exposure information and incomplete details of the effects reported to have occurred. More often than not, prediction of the toxic effects which will occur in people, for given exposure conditions, depends on data derived from animal tests.

Toxicological studies fall into two main types. Regulatory or screening tests, sometimes referred to as 'routine' studies, comprise the first type and are designed to investigate the various aspects of toxicity which are capable of being induced by chemicals. Owing to their relatively well-defined designs and selected end-points, many of these studies form the basis of the various classification systems established by regulatory bodies. The other type of toxicological experimentation embraces the more detailed, customized, investigative work into mechanisms of toxicity, often aimed at explaining the findings of studies of the first type.

For industrial chemicals, the majority of the available data stem from the first type of tests and these can be further divided and grouped according to the types of effects which they investigate. The following are the principal categories concerned.

4.1 Acute toxicity

The term 'acute toxicity' is often mistakenly equated with the LD_{50} test† and only concerned about a chemical's ability to kill. In fact, the term applies to *all* adverse effects which are observed soon after exposure, often developing rapidly and sometimes resolving quickly once exposure ceases. Generally, but not always, such effects are observed only following exposures to high concentrations of a chemical. The range of effects which can qualify as acute toxicity includes irritation responses, (e.g. skin or eye irritancy), effects on major organs (e.g. brain and depression of the central nervous system (CNS), liver damage, etc.) and, in the extreme, death. In assessing the hazard of a chemical in terms of its potential to affect adversely human health, it is less important to know that exposure to a certain dose can result in death than to know how and why the chemical exerts its effects. Thus, a good study of acute toxicity in animals will include observations of appearance and behaviour, tests of organ

---

†The LD_{50} test is a test in which groups of animals (usually rats or mice) are administered a range of single, often high, doses of a chemical so as to allow a dose to be calculated which, when administered to a group of animals, will result in the death of 50% (hence, LD_{50} = Lethal Dose 50%).

function, measurement of blood and urine constituents as well as microscopic examination of major organs after death. If deaths occur as a result of treatment, the LD_{50} (or LC_{50} in the case of inhalation toxicity) may be calculated. Such values are frequently cited in isolation under the heading 'Acute toxicity' and used for classification purposes under regulations designed to rank chemicals according to acute toxic hazard (e.g. the EC Dangerous Substances Directive, the International Maritime Organization (IMO) rules on the transportation of dangerous goods, etc.).

The irritancy and corrosivity of chemicals are important indices of acute toxicity which need to be evaluated in terms of workplace risks. Physico-chemical data may provide sufficient insight into these important properties to enable a judgement to be reached without the need to resort to animal studies (e.g. knowledge of the base reserve in the case of detergent type substances, strong acids and alkalis, etc.). Where studies in animals are judged necessary, in the case of skin irritancy and eye irritancy, specific toxicity tests are used. In the case of irritancy resulting from exposure by inhalation, information can be obtained from observations made in the 'general' acute toxicity study.

4.2 Sub-acute toxicity

Sometimes called 'short-term' or 'sub-chronic toxicity', this term includes adverse effects which develop more gradually than those described under acute toxicity. Generally, these effects result from repeated or prolonged exposures. In some cases, they may be progressive, in others, adaptation and tolerance can develop. Some effects are permanent, others reversible—although recovery may well be slow.

Good sub-acute toxicity studies involve extensive observation of the clinical condition, behaviour and growth of the animals throughout the study. The duration of exposure is usually in the range of 28–90 days in rodents. Measurement of selected constituents and parameters of the blood and urine are made during the study to monitor the health of the animals and their organ systems. Finally, following death, the organs and tissues are examined in detail to study the nature of, and possibly the reason for, any adverse effects attributable to exposure to the test substance. Such studies are designed to investigate the effects of a range of dose levels or exposure concentrations of the chemical. These dose levels are selected to define the nature of the toxic effects and establish a 'no-effect level'. Data which permit the definition of a dose-response curve are particularly useful in the hazard assessment process and in providing the basis for workplace controls of exposure.

4.3 Chronic toxicity

This term, sometimes called 'long-term toxicity', covers adverse effects which develop through repeated exposure to a chemical over a prolonged period of time (months or years). As with sub-acute toxicity, the effects observed may be progressive and permanent; others may be reversible upon cessation of exposure although this is likely to take considerable time.

Good chronic toxicity studies require similar observations to be made to those outlined for sub-acute studies. However, since the duration of exposure covers the life-span for most of the animals, there is a need to take into account, in the design of the study and its interpretation, the effects of the ageing process on the indices and parameters measured.

Such studies are designed to investigate the effects of a range of dose levels or exposure concentrations in order to define the nature of the toxic effects and establish a no-effect level. Since the studies often cover the whole life-span of the test animal, the data can be particularly useful in establishing workplace controls for substances to which there are likely to be long-term, occupational exposures for some individuals. The studies may also provide data for the determination of carcinogenic hazards (see below).

4.4 Skin sensitization
These tests are designed to predict the ability of a chemical to cause skin sensitization (also known as allergic contact dermatitis) in people. Once 'sensitized' to a particular chemical, a susceptible person may experience severe reactions following subsequent exposure to extremely low levels of the material. The most common and accepted screening methods involve testing in guinea-pigs and require specialist interpretation.

4.5 Carcinogenicity
A chemical which can induce cancer in animals is known as a carcinogen. Hence, the carcinogenicity of a substance is a measure of its ability to induce cancer. Only around thirty chemicals have been shown conclusively to be human carcinogens although many hundreds have been shown to produce cancer in other species of animal. Considerable research into this extremely important area of toxicity has shown that, like other forms of toxicity, there can be major differences of sensitivity and specificity between species, even to the extent that a substance which produces an effect in one species, may not produce the effect *at all* in another. Nevertheless, when a chemical has been shown to be carcinogenic in animals, it is prudent to assume to that it will present a similar hazard to people unless there are data which show the findings in animals to be irrelevant to people.

Carcinogenicity tests are broadly similar in design to chronic toxicity tests in that provision is made for exposure to the chemical for most of the life-span of the test species. Owing to their costs in animals, time and money, a number of cancer screening tests have been developed to assist in assigning priorities for testing; many of these tests are based on screening for mutagenic activity (see section 4.6 below).

4.6 Mutagenicity
A chemical which can induce heritable changes in an organism is called a mutagen. Up to now, no chemical has been shown conclusively to be a human mutagen. However, mutagenicity tests can provide information on the ability of a substance to damage genetic material in cells. Damage to DNA can result in cell death or cancer as well as heritable changes, depending on the type of cell affected and the nature of the damage. Numerous test systems have been developed to screen chemicals for mutagenic activity using bacterial cells (eg. Ames tests), mammalian cells in culture and whole animals.

The data from such studies require specialist interpretation and may provide useful input in the hazard assessment process.

4.7 Reproductive toxicity
Any adverse effects induced by a chemical on the process of reproduction is termed reproductive toxicity. Generally, tests here are designed to reveal adverse effects on male and female fertility, on the development of the offspring in the womb and on the development of the offspring into healthy, sexually mature adults. The most commonly used tests are:

(a) **The teratogenicity study**—which examines effects on the offspring resulting from exposure of the mother to the chemical during early pregnancy and
(b) **The reproduction study** (fertility study, single- or multi-generation study)—which examines effects on fertility and post-natal development.

All reproductive toxicity tests require considerable expertise in their design, conduct and, above all, interpretation.

5. EVALUATION OF HAZARDS

In stage four of the hazard assessment procedure, the potential hazards to health presented by the chemicals are evaluated; this forms the basis of an assessment of the risks to health of those occupationally exposed. This final stage in the process is based on the development of a clear understanding of the toxicity of the substance in relation to the exposure characteristics which are likely to be encountered. This simple statement conceals numerous difficulties and pitfalls which await even specialist practitioners in industrial toxicology, occupational hygiene and medicine. The main areas of difficulty are discussed below.

5.1 Extrapolation of animal data to people
The basis of prediction of adverse effects in people from studies of the toxicity in animals relies upon the fact that there are many similarities in the make-up and functions of test animals compared with humans and, indeed, in the manner in which they react to poisons. Even so, differences in both qualitative and quantitative responses to certain chemicals are quite common. Thus, it should never be assumed that people will react in the same way as animals—or indeed vice versa. In practice, the extrapolation process and predictions based on it require considerable specialist experience taking into account the specific toxicity information, knowledge of the reactions of animals and people to substances of similar chemical structure and, above all, a sound understanding of the normal biological process of animals and people and the manner in which these can be altered by various factors.

5.2 Mixed exposures
If the purpose of the hazard evaluation exercise is to provide the basis for assessment of risk in the workplace, it should be remembered that it is rare for the workforce to be exposed only to a single chemical and by a single route of exposure.

(a) Mixed substances

It is clearly impossible to develop toxicity data for every combination of mixed exposures likely to occur in every workplace. Nevertheless, the hazard evaluation process should consider the possibility of the following interactions occurring:

(i) *Addition*—substances acting via a similar mechanism on the same organ or organ systems.
(ii) *Independent action*—substances acting independently of each other, usually on different organs.
(iii) *Synergism*—substances acting to produce a toxic response which is great than the sum of the effects of the substances acting individually.
(iv) *Antagonism*—substances acting to produce a toxic response which is smaller than the sum of the effects of the substances acting individually.
(v) *Reaction*—substances reacting chemically to produce other substances which have different toxic properties from the original substances.

Considerable specialist expertise is needed to evaluate the above interactions, particularly (iii), (iv) and (v). Since these latter phenomena are relatively rare, for practical purposes, in controlling exposure in the workplace, it can usually be assumed that the effects of exposure to mixtures of individual substances are additive.

(b) Mixtures and preparations

Evaluation of the hazards associated with exposures to complex mixtures, in the absence of specific toxicity data, requires extrapolation from data on the component parts, taking account of any interactions that may occur. Clearly, such an approach requires detailed compositional knowledge. Where this is absent, e.g. for proprietary products (where the supplier provides only general, banded data) or products having a large number of highly variable chemical constituents which are produced to technical specifications, great care will be needed in their evaluation and some reliance will have to be placed on the supplier's advice.

(c) Mixed routes of exposure

Evaluation of workplace hazards of individual chemicals (or various mixtures thereof) should take account of the likely routes of exposure and their combined effect where more than one route is possible. For example, control of airborne levels of *n*-hexane, in order to protect against the development of nerve damage, may be ineffective if there is no control of exposure through skin contact.

5.3 Inadequacy of data

Information useful for assessing the hazardous properties of substances may be obtained from a number of sources. These include data provided by manufacturers or suppliers in their material safety data sheets or on product labels; guidance on exposure limits provided by the American Conference of Governmental Industrial Hygienists (ACGIH) and regulatory authorities (e.g. HSE's Guidance Booklet EH 40) and data from various technical reference sources and professional bodies.

Where a recommended or statutory occupational exposure limit (OEL) has been established for a given substance, its hazard assessment and control in the workplace may be straightforward. It must be remembered, however, that such OELs relate to exposure by inhalation and that there will still be a need to consider the contribution which may be made by exposure via other routes (e.g., skin absorption) to the overall risk to health (see Section 5.2(c) above).

Experimental and human toxicological data may also be available from technical reference sources, e.g. scientific research papers, textbooks, etc., or from experience gained and information generated by central and local, in-house specialists in occupational health, e.g., toxicologists, occupational hygienists and physicians.

Clearly, hazard evaluation can only be as good as the data upon which it is based. It is a frequent problem of hazard evaluation for the database to be deficient in one or more respects. The following are the more frequently encountered reasons for such deficiencies.

(a) Identity
Precise definition of the substances, including any contaminants, is a crucially important part of the evaluation since it will underpin judgements made by comparison with other known substances, as well as forming the basis of understanding of the toxicity data and their relevance to people. Where such information is deficient, greater reliance needs to be placed on suppliers' data, where available, or on other sources of information, the validity of which may be difficult to establish.

(b) Inappropriate exposure route
Whilst toxicity data may be available for a substance relating to effects arising from a variety of exposure routes, it is quite common to find information gaps for routes of particular significance to occupational exposure. For example, there may be extensive data on the oral toxicity of a substance, to underpin its use as a food additive, but few data on its toxicity by skin contact or inhalation, routes of great importance in the workplace. Considerable experience is necessary in extrapolating data from one route of exposure to another and this cannot be reliably achieved without additional specialized data, e.g., on metabolism, pharmacokinetics, etc.

(c) Inadequate duration of exposure
Data may be available which are adequate to support the prediction of effects in people resulting from a single exposure to splashes of the chemical onto the skin or into the eye but no information from which to judge the effects of long-term, repeated, low-level exposure. It the substance is used in closed systems, with effective controls to ensure that exposure will be occasional or rare (e.g. when sampling or with accidental spillage), the available data may be sufficient. On the other hand, additional information would undoubtedly be required to evaluate fully the hazards of a substance used in large quantities by many people over extended periods of time.

(d) Unclear no-effect levels

An important objective of the hazard evaluation process is to provide an estimate of the exposure level likely to be safe for people. Such an estimate is frequently based on knowledge of levels of exposure which have resulted in no untoward effects in animal studies. The magnitude of the safety factor to be used in deriving the workplace exposure limits varies according to the seriousness of the toxic effects and the quality of the data. Where the data are unclear, making the establishment of a no-effect level uncertain, it is impossible to derive other than provisional limits for the workplace, pending development or collection of further data (See also Chapter 10).

6. PERIODIC REVIEW

The hazard assessment process should not be seen as a once-and-for-all exercise. Periodically, there will be a need to undertake a review of the hazard assessments in order to take account of any new data which become available, e.g. from the results of health surveillance, whenever new or modified plant, processes or work practices are introduced which might give rise to different patterns or degrees of exposure and, finally, at minimum regular intervals, e.g. every 5 years, in order to ensure updating of the assessment to meet continuously evolving standards.

7. CONCLUSIONS

Assessment of the nature and severity of the toxic hazards associated with exposure to chemicals is fundamental to the development of appropriate controls to protect the health of workers. In addition, the process of hazard assessment provides the basis for the following:

(a) preparation of product safety data sheets,
(b) classification according to the criteria of the Classification, Packaging and Labelling Regulations 1984, (and equivalent),
(c) establishment of occupational exposure limits,
(d) development of first aid measures for accidental exposures,
(e) formulation of advice on medical treatment for over-exposures,
(f) information to be provided to poisons centres, and
(g) development of methods for biological monitoring.

Hazard assessment is a complex process often requiring input from occupational health professionals. It should be repeated at regular intervals to take account of developing standards and knowledge as well as when changes in usage pattern occur or new data become available.

7

Assessment of health risks

Mr J. Holding and Mr N. Budworth

SUMMARY

Under Regulation 6 of the Control of Substances Hazardous to Health Regulations (COSHH) 1988, there is a requirement to assess the health risks created by work involving substances hazardous to health.

This chapter is designed to inform the reader of the substances (or classes of substances, including micro-organisms) to which employees may be exposed and the effects these substances can have on the body. Guidance on sources of information and its interpretation is also included.

Because of the very wide scope of COSHH, it is impossible to cover all substances except in certain specific cases; the aim is to give broad-based guidelines relating to groups of substances covered by COSHH. Examples are given on how Bayer PLC has tackled the requirements of the regulations. These relate to process manufacture, research laboratories, warehousing and distribution.

The chapter also looks into methods of record-keeping including hard copy and computerized systems.

1. INTRODUCTION

Although the assessment of risks has been an area of increasing importance in safety management in recent years, it is the Control of Substances Hazardous to Health Regulations which introduced a legal duty for the formal assessment of risks. Several other sets of regulations are now following this approach but it is COSHH which laid down the groundwork.

Regulation 6 of COSHH states that 'an employer shall not carry on any work which is liable to expose any employees to any substance hazardous to health unless he has made a suitable and sufficient assessment of the risks created by that work to the health of those employees and the steps that need to be taken to meet the requirements of these regulations'.

'The assessment required by the above paragraph shall be reviewed forthwith if

(a) there is a reason to suspect that the assessment is no longer valid; or
(b) there has been a significant change in the work to which the assessment relates,

and where, as a result of the review, changes in the assessment are required, those changes shall be made'.

This chapter aims to differentiate between the terms 'hazard' and 'risk' and give practical examples of how a large chemical company tackles the relatively new duty to carry out risk assessments.

The Health and Safety Executive (HSE) leaflet, 'Hazard and Risk Explained', [1] gives the following definitions:

The *HAZARD* presented by a substance is its potential to cause harm.
The *RISK* from a substance is the likelihood that it will harm you in the actual circumstances of use.

These concepts are illustrated by the following practical example. Asbestos is a very hazardous substance which has the potential to cause serious, long-term, lung damage. If the asbestos is kept within a sealed container, the risk is extremely low as no-one can come into contact with it. The risk only becomes significant if the asbestos is removed from the container.

Proper carrying out of COSHH assessments depends on a clear understanding of the difference between the hazard and the risk. In fact, HSE's booklet, 'COSHH Assessments', [2] states that the duty of assessments involves:

> 'Evaluating the risks to health arising from work involving substances hazardous to health, and then establishing what has to be done to meet the requirements of the whole of the COSHH regulations'.

2. TYPES OF ASSESSMENT

There are essentially four ways of carrying out assessments, based on substances, processes or areas, controls, or done generically. Each type of assessment has some similarities and differences which are discussed below.

2.1 Substance-based assessments

Substance-based assessments are carried out where a particular substance is chosen and its risks to health assessed at each stage of its use. This type of assessment is only practicable where relatively few substances are used, as the paperwork involved can be considerable.

It is used only where the substance is particularly hazardous and where its use needs a critical review.

If this type of assessment is to be used, each point at which exposure to the substance can occur should be considered from delivery to disposal. Each chemical interaction or change of state must be given due consideration together with the possibility of accidental release or exposure.

2.2 Process (or area)-based assessments

The process- or area-based assessment is probably the most commonly used type of assessment. In this case, a particular process is selected and assessed. The size of the process (or area) for each assessment made depends on the risk or complexity of the process.

If a particularly hazardous or complex process is to be assessed, the area of the assessment should be kept small, perhaps just a part of the overall process, for example, the loading of raw materials or the use of a degreasing bath.

If the risk of the substance is believed to be low, a larger area can be considered. For example, it is quite common to consider a kitchen or a cleaner's cupboard in one assessment.

If an area- or process-based assessment is to be used, care must be taken to ensure that the assessment takes into account the way the materials are actually used and not the way they are supposed to be used.

2.3 Control-based assessments

The other two types of assessment are mainly used for special cases.

Control-based assessments are generally used in laboratories where several fixed control measures are usually in place but where a large number of possible processes can be in operation. Laboratories usually involve competent people who can be expected to perform their own risk assessments with some additional training. It is to this end that we have produced the form reproduced in Fig. 1. Using this form, research chemists can perform simple tick-box assessments before they start each experiment. This can then be attached to the experimental log to give a complete record of the exposure.

The Royal Society of Chemistry (RSC) has described a similar approach in its booklet, 'COSHH in Laboratories' [3]. However, this involves a risk assessment matrix which, for our particular application, is difficult to apply but which has proved invaluable in other applications.

Requiring trained research chemists to perform risk assessments for each experiment they perform may initially seem to be an arduous and time-consuming undertaking. However, this procedure helps to concentrate the mind and makes people aware of the hazards of the materials they are about to handle.

We intend, in our company in the near future, to integrate an exposure logging function with the chemicals inventory and chemical location system. This will mean that the items and frequency of use can easily be determined for future assessments and epidemiological studies.

108 Assessment of health risks [Ch. 7

Experiment no(s)		Toxic	Very toxic	Harmful	Corrosive	Irritant	Explosive	Oxidizing	Flammable	Health hazards*
Materials (reagents (R)) (by-products) (products (P))										

SAFETY PRECAUTIONS

☐ Fume hood ☐ Gloves

☐ Face shield ☐ Safety shield

Make colleagues aware

Safety spectacles/goggles

SPECIAL SAFETY/CONTROL MEASURES

e.g. breathing apparatus, colleague standing by with a fire extinguisher

POST-REACTION COMMENTS

*Note: entries in this column will require the use of control measures

Experimenter .... Supervisor ....

Fig. 1. COSHH risk assessment form.

2.4 Generic assessments

When performing generic assessment, particular care must be taken. They can be classed as either process- or substance-based.

In generic, process-based assessments, a simple process is assessed. An example is the use of a photocopier, the results of which can then be applied to other photocopiers. However, care must be taken to ensure that the other cases are, indeed, similar and that the assessment does, therefore, apply. A photocopier in a small, poorly ventilated room may require a slightly more specific assessment than one in a large open-plan office. However, if the worst case is assessed, i.e., a photocopier in a small, unventilated room, and similar control measures applied to another photocopier, say in a large office, the latter will be adequately controlled. Fairly frequent reviews are also required to ensure that the assessment conditions are still valid.

This type of assessment has obvious advantages. For example, a chain of shops need not be assessed individually, thus avoiding much duplication of effort. If the conditions in a particular location are not exactly similar, enough information should be found in the generic assessment to allow the local COSHH assessor to make a valid assessment without repeating too much of the groundwork.

Generic, substance-based assessments are similar to the above in that similar items are grouped together for the sake of the assessment. In this case, however, substances are grouped. This is an approach that we have used successfully in the flavours and fragrances industry. Substances are grouped according to their risks and routes of entry into the body. For example, harmful hydrocarbon solvents may form one group and corrosive liquids another. Care must be taken, however, to ensure that materials are correctly grouped. A particularly obnoxious material may well be worth considering as a 'group' of its own and paying special attention to it. These groups are then assessed in the usual way with the worst material in each group being selected as the basis for the assessment. This assumes that, if the worst case is properly controlled, so will all the other cases.

Deciding which method of assessment to use is, of course, only the first step. Whichever method is chosen, the depth and detail required will depend on the risk and complexity of the operation.

In the most simple cases, assessments do not have to be recorded. However, we recommend that all assessments, however simple, should be documented, i.e., where the result of an assessment is obvious and can easily be repeated. An example of this might be the use of correction fluid in a typical office situation. (See Case Example (a) in section 6 below).

More hazardous processes require more detailed written assessments. Processes, such as routine cleaning or those in small engineering workshops require assessments of this sort. Details of data needed for this type of assessment are shown in Fig. 2(a). These are the most frequently performed assessments and do not require excessive detail. However, information for this type of assessment must still be gathered from a variety of sources and it is this that takes the time with assessments. The more comprehensive the information, the more reliable the assessment and the more certain one can be of the conclusions. This means that every minute spent on obtaining information will provide benefits later.

(a) A Routine Assessment	(b) A Detailed Assessment
— Area involved — Operation — Type of work activity — List of substances involved which are hazardous to health — List of substances with an Occupational Exposure Standard or Maximum Exposure Limit (with their limits) — Substances which have a hazardous decomposition product — Substances which have hazardous reactions/reaction products — Substances which have synergistic reactions — Hazards and possible effects associated with these products — Possible routes of exposure — Likelihood of exposure—extent of deviation and people involved — Previously noticed effects — Whether monitoring is required and results of previous monitoring — Whether health surveillance is required — Conclusions on the risks to health — Possibilities of elimination or substitution — Name and signature of assessors — Date (Safety data sheets are attached at the end of the assessment)	— Site — Area involved — Operation — Number of workers exposed — List of substances used (Name, physical state, mode of exposure, toxicity class) — Occupational exposure limits — Sketch or flowchart of process — Sources of exposure during storage and, possibly of, leaks — Suitability of packaging and sources of exposure during transport, transfer and possibility of spills — Details of use and details of exposure during use — Disposal of excess material and possibility of exposure — Emissions to atmosphere — Waste products — Intermediates (at each stage, the possibility of exposure by various routes of entry into the body is detailed along with methods for control) — Monitoring details of previous surveys — Health surveillance; relevant facts from previous surveys — Control measures required — Biological monitoring, details of previous survey — Details of training given/needed — Welfare and personal hygiene facilities — Copies of health and safety work-sheets and material safety data sheets — A statement saying whether a hazard to health is present — Corrective action — Name (Signature) — Date (and date of next assessment)

Fig. 2. Data required to carry out assessments.

Processes of a high complexity or posing a high perceived risk need to be assessed in far more detail. Similarly, where particularly hazardous or carcinogenic substances are used, the level of detail in the assessment needs to be high. For example, our assessment for the unloading of acrylonitrile from road tankers runs to more than ten pages. Details of the data we have used for these situations are given in Fig. 2(b). In such cases, the assessment closely resembles a hazard and operability study (HAZOP), examining every possible point of exposure and determining ways of dealing with a loss of containment while protecting the workforce.

The above text has discussed various types of assessment without actually giving a simple definition of the assessment. In our opinion, the assessment should be a summary of all available information with a judgement of the level of risk to health and a statement as to the control measures required.

A significant amount of the information required for an assessment may already be available in standard operating procedures, laboratory method sheets or quality assurance documentation. We have tried to reduce the workload in our organization by adding the additional details to these documents and avoiding duplication of the existing portion.

3. EXPOSURE MONITORING

When COSHH was introduced, many companies initiated major programmes of exposure monitoring so that they would have a complete picture of data for the assessment. With the benefit of hindsight, however, it can be seen that much of this monitoring was unnecessary. First, because, in some cases, the level of exposure was either obviously extremely high or insignificant. Secondly, the monitoring may well have been performed for a similar situation before. In such cases, it is advisable to contact trade associations and suppliers before embarking on a monitoring programme.

The first step in any monitoring programme should be to decide what is to be achieved and whether monitoring will help in this. Monitoring is not covered in detail here as it is the subject of Chapter 11 of this book. Anyone planning to initiate a programme of monitoring should also consult HSE's Guidance Note EH 42 on 'Monitoring Strategies for Toxic Substances' [4].

4. COLLECTION OF INFORMATION

To perform any type of assessment, information is required. The quality of the assessment depends greatly on the quality and extent of the information obtained for it. Obtaining the required information can be a simple or a very difficult task. Some useful sources of information and advice for this are discussed below.

4.1 Material safety data sheets

The first place to turn to for information is the suppliers of the materials involved. They have a duty under Section 6 of the Health and Safety at Work Act (as amended) to supply information to allow the material to be used safely. This information is usually in the form of a material safety data sheet. Fig. 3 lists the data items which should be included in material safety data sheets.

- Product name
- Intended uses
- Composition
- Physical and chemical properties
- Health hazards
- Fire hazards
- Storage precautions
- Transport precautions
- Handling and use
- Precautions (including personal protective equipment advice)
- Emergency action (fire, spillage and first aid)
- Disposal precautions
- Additional information
   (ecological hazards, relevant regulations, toxicological data references)
- Name, address and telephone number of the supplier
- Reference number
- Date of issue

Fig. 3. Data which should be included in material safety data sheets

An example of a typical safety data sheet for a product supplied by Bayer is reproduced in Fig. 4. Within the next few years, all such data sheets in the European Community should be produced in a standard format. However, the main problem with safety data sheets is that they vary greatly in quality; some are well researched and contain excellent information whilst others are devoid of any useful information. The first task must be to decide whether the data sheet contains enough useful information for an assessment to be made. If it does not, the next step should be to refer back to the manufacturer and ask for further details. It should be emphasized though that a standard data sheet has not been developed for COSHH assessments so that the data sheet should be regarded as only one source of information to be tapped when carrying out an assessment.

If the additional enquiry does not produce enough information to enable a suitable and sufficient assessment to be made, other information sources need to be consulted.

4.2 HSE literature
The next place to search for information is in the HSE literature. Each year, the Executive publishes a large volume of literature and has compiled a COSHH subject catalogue listing all their publications relevant to the COSHH Regulations [5].

4.3 Trade associations
If, after consulting this, further information is still needed, the next place to look is probably a trade association. Trade associations are knowledgeable on their product areas and can help to direct enquiries to suitable sources of information. Some trade associations also produce their own information lists. For example, the British Rubber Manufacturers' Association (BRMA) has produced a book entitled 'Toxicity and Safe Handling of Rubber Chemicals' [6]. This contains information on many chemicals

Inorganic Chemicals Group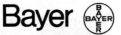

DIN Safety Data Sheet 000020/01
Date of issue: January 11, 1991   Page 01 of 03

Company Bayer PLC
         Bayer House, Strawberry Hill, Newbury, Berkshire RG13 1JA
         Tel.: (0635) 39000

Commercial product name Caustic soda (Pearls/Flakes)

1.1 Chemical characterisation: Sodium hydroxide (approx. 99 % NaOH)
                               CAS–No.: 1310–73–2
1.2 Form: solid
1.3 Colour: white
1.4 Odour: odourless

2.   **Physical and safety data**   tested in accordance with
2.1 Change in physical state:
       Melting point: 319 °C
       Boiling point: 1390 °C at 1013 mbar
2.2 Density: 2,13 g/cm^3 at 20 °C
2.3 Vapour pressure: not applicable
2.4 Viscosity: not applicable
2.5 Solubility in water: 1260 g/l at 20 °C
                                 3410 g/l at 100 °C
2.6 pH value: > 14 at 100 g/l water at 20 °C
2.7 Flash point: not applicable
2.8 Ignition temperature: not applicable
2.9 Explosive limits: not applicable
2.10 Thermal decomposition: not applicable
2.11 Hazardous decomposition products: No hazardous decomposition products
                                 observed.
2.12 Hazardous reactions: Reacts violently with water, acids and ignoble
                                 metals as aluminum, magnesium, zinc
                                 (H_2–development).
2.13 Further information: hygroscopic

3.   **Transport**
     GGVSee/IMDG Code: 8   UN No.: 1823   MFAG: 705 EmS: 8   06
     GGVE/GGVS: Class 8   No. 41B   RID/ADR: Class 8   No. 41B
     ADNR: Class 8   No. 31A   Cat — ICAO/IATA–DGR: 8   1823 II
     Postal dispatch approved: no
     Declaration for land shipment: NATRIUMHYDROXID (FEST)
     Declaration for sea shipment: Sodium hydroxide, solid
     Other information:
     Corrosive. Harmful. Keep dry. Keep separated from foodstuffs.

DIN 52900   (GB) 000002

Fig. 4. Material safety data sheet for caustic soda.

Inorganic Chemicals Group

DIN Safety Data Sheet

Date of issue: January 11, 1991

000020/01

Commercial product name Caustic soda (Pearls/Flakes)

4. Regulations

Labelling according to UK Classification Packaging and Labelling of Dangerous Substances Regulations SI 1984 No 1244 and corresponding EEC directives:
Symbol: C, hazard description: corrosive
anhydrous sodium hydroxide (EEC №. 011-002-00-6)

R 35:	Causes severe burns.
S 2:	Keep out of reach of children.
S 26:	In case of contact with eyes, rinse immediately with plenty of water and seek medical advice.
S 37/39:	Wear suitable gloves and eye/face protection.

Occupational Exposure Standard: 2 mg/m^3

5. Protective measures, storage and handling

5.1 Technical protective measures: Keep container tightly closed and dry.
Handle and open container with care.
Never add water to this product.

5.2 Personal protective equipment: Eye protection: Wear eye/face protection.
Hand protection: of rubber
Respiratory protection: Dust-protection mask if there is a risk of dust formation.

5.3 Industrial hygiene: Avoid contact with eyes and skin.

5.4 Protection against fire and explosion: No special measures required.

5.5 Disposal: Must be given special treatment on consultation with the manufacturer.
Dispose of empty packages in an official refuse dump.

6. Measures in case of accidents and fires

6.1 After spillage/leakage/gas leakage: Take up avoiding formation of dust.

6.2 Extinguishing media: No restriction in fire situations. (see capitel 2.5 and 7)

6.3 First aid: Take off immediately all contaminated clothing.
After contact with skin, wash immediately with plenty of water.
Contamination of the eyes must be treated by thorough irrigation with water, with the eyelids held open. A doctor (or eye specialist) should be consulted immediately.
If product is swallowed immediately drink a lot of water/milk. Do not induce vomiting. Medical treatment is required as soon as possible.

6.4 Further information:

DIN 52900   (GB) 000002

Fig. 4. (*Continued*).

Inorganic Chemicals Group

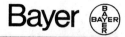

DIN Safety Data Sheet	000020/01
Date of issue: January 11, 1991	Page 03 of 03

Commercial product name   Caustic soda (Pearls/Flakes)

7. **Information on toxicity**
   Effect on the eyes: severely corrosive
   Effect on the skin: severely corrosive

8. **Information on ecological effects**
   Fish toxicity: LC_{50}: 189 mg/l
   (Roth,L.; Wassergefährdende Stoffe)
   Toxic effect on fish, plankton and on sedentary organisms, also through shifting of pH value.
   Causes no biological oxygen consumption.
   Behaviour in wastewater treatment plants: No inhibition of activity of waste water bacteria after neutralisation.
   Hazard class (WGK): 1 — slightly hazardous to water
   WGK = Classification in accordance with the West German Water Resources Act

9. **Further information**

The data given here is based on current knowledge and experience. The purpose of this Safety Data Sheet is to describe the products in terms of their safety requirements. The data does not signify any warranty with regard to the products' properties.

Fig. 4. (*Continued*).

commonly used in rubber processing in sufficient detail to perform an assessment. The Chemical Industries Association (CIA) has produced a series of publications on COSHH covering all aspects of the regulations. (See Chapter 1 and Appendix II at the end of the book).

4.4 Other literature sources

If enquiries within trade associations prove unsatisfactory, the next avenue to consider should be literature sources. Many books are available giving information on toxicity and hazards in use. Probably the most common are 'Sax's Dangerous Properties of Industrial Materials' [7] and 'The Merck Index' [8]. The Merck Company Ltd has produced a comprehensive book on hazard data sheets [9]. Any or all of these should provide useful information on the hazards of particular substances.

If no information can be found from the above sources, the final step is 'on-line', computer searching, i.e., a search of the toxicological databases of the world. This can be carried out through a host such as 'Datastar' (actually on-line) or an approach to a commercial company to conduct the search. Increasingly, toxicological databases, such as 'Toxline', are becoming available on CD-ROM and are therefore becoming more accessible.

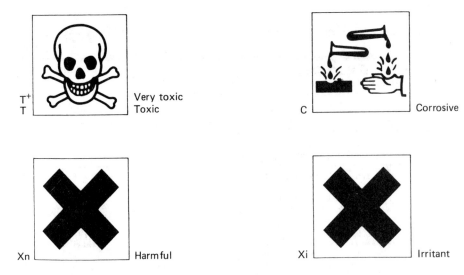

Fig. 5. Hazard warning labels.

One source of information which has not yet been mentioned (in this chapter) but which is usually the most accessible and the easiest to understand is the product label. Hazardous substances are required to carry a label, as specified by the Classification, Packaging and Labelling of Dangerous Substances Regulations 1984 [10]. The label consists of a pictogram, a general indication of the risk (i.e., very toxic, toxic, harmful, etc.) and a set of standard risk and safety phrases. This is shown in Fig. 5. The risk and safety phrases are uniform throughout the EC so that, if the label appears in a different language, the meaning of the text can be found merely by referencing the phrase number.

These labels give a simple and immediate indication of the hazards possessed by the material. Once the information has been gathered, the actual assessment can begin.

Finally, for a fuller discussion on evaluation of hazards, the reader should refer to Chapter 6.

5. EVALUATION OF THE RISKS TO HEALTH

From the data obtained, the hazards to health and the routes by which the chemicals can produce harmful effects can then be considered.

There are only a limited number of routes by which chemicals can enter the body (see Fig. 6). A material only causes harm if it is present in a state in which it can enter the body, by one or more routes, and, if it has an effect on the body, by being absorbed by that route. For example, metal in the form of blocks is unlikely to be harmful. However, in the form of fine dust or fume, it can be inhaled and can therefore be a hazard to health. Similarly, some substances can be absorbed through the skin while others cannot. This means that an important part of the assessment should involve looking at places where exposures can occur, the forms in which the substances are present and the routes of entry into the body open to them.

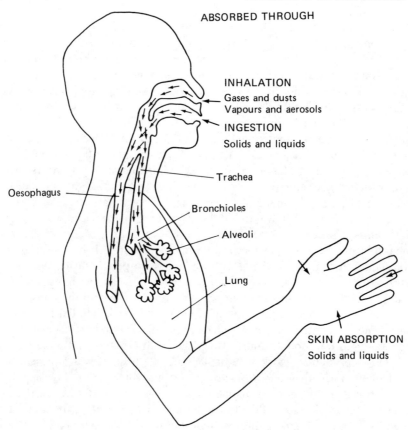

Fig. 6. Routes by which chemicals can enter the body.

For example, a person may be exposed to a substance which can be very toxic if ingested but which may not be harmful if its only contact is with the skin. This assumes that the standard of personal hygiene is such that the substance is removed from the skin by cleaning before any is likely to be ingested. This situation is mentioned here solely for illustrative purposes and should not be allowed to occur in practice. Particular attention must be paid to substances which can cause harm by inhalation or by absorption through the skin as these are the most likely routes actually to cause harm.

It is extremely difficult to measure the amount of a material which can be, or has been, absorbed through the skin. So, in practice, if a material can cause harm by skin absorption, no skin contact should be allowed.

Part of the process of determining where a risk is likely to arise consists of examining previous monitoring data (either airborne or biological data) and using techniques, such as dust lamps, to see whether there is a problem. As previously mentioned, in cases where it is obvious that exposures are extremely high or low, monitoring is pointless unless it is impracticable to reduce the levels of the contaminant and the levels are required so that the most appropriate protective devices can be chosen.

Much useful information can be gained by speaking to people on the shop floor. They will identify a process if it is particularly smelly or dusty or one with which they are unhappy. Finally, particularly in larger companies, it is worth examining absenteeism records and any cases of over-exposure which may have occurred. These can give a valuable insight into any effects which the exposures may be having.

If the assessment relates to a machine or operation, it is important to review the whole cycle of the operation in conjunction with what the operator of the process considers to be the worst material(s) involved. For example, we found that exposures during the bagging of roast chicken-flavoured powder, at our flavours and fragrances plant, were significantly higher than those with most other flavours. Moreover, exposures were higher still at the start of the filling run than at the end, due to minor adjustments having to be made to the local exhaust ventilation at the start of a run. Similarly, in an oil factory, it was found that exposures to oil mists were only significant in the filling-off procedure when lines were blown clear at the end of a run. This action produces a fine oil mist which is not present during other parts of the procedure.

Consideration should also be given to people who perform each task. It is a fact of nature that one operator may be able to perform a task without spilling a drop of liquid or without creating any dust while another will be bathing in the liquid and engulfed in a dust cloud. Individual differences must be accounted for and working arrangements or controls recommended to suit the worst case. When considering sources of exposure, thought must also be given to cleaners and maintenance staff. The latter, in particular, work in situations where substances are present at higher levels than are normally found anywhere else in the workplace and these people must be fully protected. In addition, maintenance staff may be less aware than process operators of the risks presented by the materials in use. The assessment may therefore be used as a way of familiarizing such workers with the hazards. In any case, the assessment must make clear that these people should be informed of the risks.

In a similar vein, the assessment must take into account what may foreseeably happen in the event of accidents. One company has already been taken to court for failing adequately to assess the hazard to health of circuit boards over-heating in an oven. Other examples of foreseeable hazards include the mixing of chemicals or the formation of chemical intermediates.

One example where we have had a great deal of experience is with COSHH in warehouses, an area where, on the whole, COSHH is quite simple to apply. The hazards of a substance may be high (i.e., it may be toxic or very toxic) although the risks are low because the substance is completely contained. The risk only becomes significant if the containment is breached, e.g. a drum is punctured. This is a totally foreseeable event and so must be planned for and an assessment made. It may be that the only practical measure is a good procedure for dealing with spillages with appropriate personal protective equipment provided for people needing to clean up spills; however, the assessment must recognize this.

6. CONTROL MEASURES

Once the risks of the operation have been evaluated, the appropriate control measures can be determined. A hierarchy of control measures is listed in the general Approved Code of Practice (Paragraph 34) to the COSHH Regulations [11]. This list is reproduced in Fig. 7.

(a) *For preventing exposure*
- (i) Elimination of the use of the substance
- (ii) Substitution by a less hazardous substance or by the same substance in a less hazardous form

(b) *For controlling exposure*
- (i) Totally enclosed process and handling systems
- (ii) Plant or processes or systems of work which minimize the generation of, or suppress or contain, the hazardous dust, fume, micro-organisms, etc., and which limit the area of contamination in the event of spills and leaks
- (iii) Partial enclosure, with local exhaust ventilation
- (iv) Local exhaust ventilation
- (v) Sufficient general ventilation
- (vi) Reduction in number of employees exposed and exclusions of non-essential access
- (vii) Reduction in the period of exposure for employees
- (viii) Regular cleaning of contamination from, or disinfection of, walls, surfaces, etc.
- (ix) Provision of means for safe storage and disposal of substances hazardous to health
- (x) Suitable personal protective equipment
- (xi) Prohibition of eating, drinking, smoking, etc. in contaminated areas
- (xii) Provision of adequate facilities for washing, changing and storage of clothing, including arrangements for laundering contaminated clothing

Fig. 7. Measures for preventing or controlling exposure. (These can be any combination of the above.)

Assessment of risks is a subjective judgement, in many ways, and, as such, cannot be performed by machines or computer programmes. In different circumstances, a given substance can give rise to different degrees of risk.

Case examples
The following examples is given, illustrating two distinctly different use situations and corresponding different levels of risk for a given chemical.

Case (a)
If an assessment is carried out on a well-known typing correction fluid, containing 1,1,1-trichloroethane, it will be found that it is labelled 'harmful' with the St Andrew's Cross. If a supplier's material safety data sheet had not been obtainable, enough information is available from the label to perform a basic risk assessment. The label

indicates that the material is 'harmful' by inhalation and ingestion. If only the chemical name had appeared on the bottle, it would still have been possible to evaluate the hazard by referring to the Authorised and Approved List. On questioning the office staff, it was found that the material was used occasionally, in 30 ml bottles, and always in a well-ventilated office. From this, it can be concluded that the only risk is from deliberate misuse.

Case (b)

The same substance, 1,1,1-trichloroethane, is also used as an industrial degreasing agent, frequently in degreasing baths and sprays. In a degreasing bath, the substance can be present in quantities of several litres. Because of the large volumes involved and the fact that the bath is frequently used, the situation must be further investigated. It is found that 1,1,1-trichloroethane has a Maximum Exposure Limit (see the Appendix at the end of this chapter); this suggests that care must be taken with this substance as MELs are set for only a few special substances. With this kind of limit comes the duty to reduce exposures to as far below the MEL as is reasonably practicable. Current values for all exposure limits adopted officially for the UK are to be found in HSE's Guidance Booklet EH 40 [12].

Further investigations reveal that the liquid can have a degreasing effect on the skin and, in cases of frequent or prolonged contact, can give rise to dermatitis. On inspection of the degreasing bath, it is noticed that one of the operators is using the trichloroethane to remove oil from his hands, a practice not previously documented or noted.

Because of the dermatitis risk and COSHH, the assessment must state that this process of cleaning hands with trichloroethane must not occur and that alternative cleaners, possibly combined with barrier and/or reconditioning creams, must be supplied. Returning to the degreasing bath, the operators say that, when they are removing material, there is a strong smell of trichloroethane. On consulting the manufacturer, it is discovered that the bath is fitted with lip extraction and fail-safe cooling systems so that it shuts down if the cooling mechanism fails. In further talks with the operators, it is concluded that, under normal operating conditions, atmospheric levels of trichloroethane should not present problems.

However, because of the operators' complaint, it is decided to perform a simple direct, colour-change atmospheric test, a simple and cheap way to determine whether the atmospheric level is high or low. The test shows the level to be close to the MEL, clearly contrary to information from the suppliers. At this point, a more accurate monitoring technique could be used to determine the precise airborne concentration or the cause of the high levels could be sought. In this case, it appears that the lift, used to raise and lower items into the bath, has been upgraded to increase productivity, the increased speed in removing items leading to the raised atmospheric levels. Thus, the assessment would have to conclude that there may be a risk to health because the levels are close to the MEL. However, the levels can easily be reduced by down-grading the motor.

This should therefore be recommended and the process reassessed when the modification has been carried out.

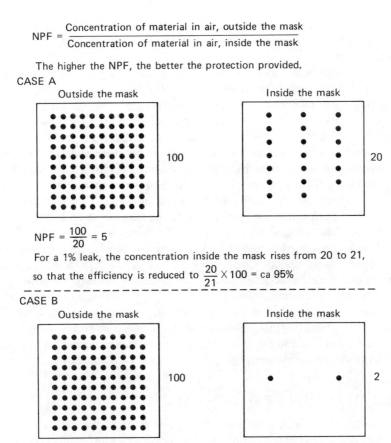

Fig. 8. Nominal Protection Factor (NPF): The effect of a poor seal on efficiency.

Thus, a check is made on what would happen if the cooling system should fail and it is decided that the arrangements in place are adequate. Periodic checks on the lip extraction must also be carried out as required by the Regulations. In addition to the above, the bath must be cleaned out at fairly frequent intervals. Consequently, risks to health must also be assessed during the emptying, cleaning and filling of the bath to ensure that appropriate control measures are in place during each of these stages.

Thus, summarizing these two situations, although the same substance is used in each case, the assessments and the risks are completely different. The level of detail and the background work required for the second case are, by necessity, far greater than that for the first. The second case also highlights an important point in that not only do reasonably foreseeable events have to be assessed but also cleaning and maintenance operations.

To ensure compliance with the remainder of the COSHH Regulations, the assessment should also recommend suitable control measures.

When considering appropriate control measures, personal protective equipment (PPE) should be viewed as the last line of defence. Initially, it may seem a cheap option but, in the fullness of time, taking account of replacement costs, the costs can mount. The level of protection afforded by some items of PPE is also debatable. For example, some items of respiratory protective equipment can leak. Frequently, leaks can occur through poor sealing around the bridge of the nose or facial stubble. Such leaks seriously affect the protection given by the mask; the greater the nominal protection factor, the greater the significance of the leak (see Fig. 8). It should also be remembered that working in these items for long periods is uncomfortable; operators frequently remove items of PPE through discomfort or so that they can complete the work more quickly. Engineering controls offer a better level of protection. Any PPE chosen must be matched to the job in hand; protection factors must be compared with operating conditions. Substances likely to be encountered must also be considered. A long way off is the universal, disposable paper mask which will protect the wearer from everything. However, judging by the actions of some operators, many believe that it is already widely available. See Chapter 9 for a fuller treatment of personal protective equipment.

7. PEOPLE WHO CAN CARRY OUT ASSESSMENTS

Having discussed the principles of assessments, one must then decide who can carry out the assessments. One of the initial consequences of COSHH was a rapid growth in the number of consultants. As always, these are of varying quality and offer varying degrees of value for money.

The HSE has been critical of some assessments performed by consultants. It should be possible to use in-house expertise and experience to perform all but the most difficult of assessments. The COSHH Regulations require that competent people are used to perform assessments. However, the level of competence required can vary with the risks and complexity of the operations. Only a little special knowledge is required to perform basic assessments, such as the correction fluid case (above). Even in more complex cases, the local manager or works chemist may have more knowledge in the way of experience with the materials and familiarity with the operating practices than is often appreciated.

COSHH is a developing process and not a once-and-for-all thing. Money spent on a consultant may be better spent on training in-house staff. However, in small companies, there may be situations where a consultant is valuable, e.g. for a specially difficult assessment or where detailed monitoring is required. If this is the case, a suitably qualified consultant should be engaged. A check should be made on the consultant's qualifications and approval (by such bodies as the Institution of Occupational Safety and Health (IOSH) or the British Occupational Hygiene Society

(BOHS)) or Membership of a specialist register (such as those maintained by the IOSH and the Royal Society of Chemistry (RSC)). When the assessment has been completed, together with any monitoring required, the real work begins. It should be remembered that the regulations are called the *Control* of Substances Hazardous to Health not the assessment of substances hazardous to health. Thus, hazards to health should be controlled, i.e., the recommendations arising from the assessment must be implemented. It should also be remembered that a COSHH assessment is not a static task; factors need to be reassessed periodically to make sure that workplace conditions have not significantly changed. Procedures also need to be in place to monitor the introduction of new processes or substances and to make sure that assessments are made *before* work starts. Items can be quarantined for quality control purposes, so why not for COSHH?

8. COMMUNICATION OF RESULTS OF ASSESSMENTS

Another important aspect is communication; the results of assessments and any monitoring undertaken must be communicated to the workforce. This is not an optional extra; it is required by Regulation 12 of COSHH as well as the general duty under the Health and Safety at Work Act. It is extremely important to convey the results of assessments in a manner which can be understood by all and in a way which neither alarms nor trivializes the risks to health—a very narrow path indeed. Assessments should be open documents and be available for inspection by the workforce, at any time. Preferably, a copy should be kept at the place of work.

9. RECORD KEEPING AND COMPUTERIZATION

As mentioned above, documentation of assessments is a legal requirement except in the most simple and easily repeatable cases. It is advantageous, in not-too detailed cases, to integrate with existing systems, for example, standard operating procedures or BS 5750 documentation. Above all, assessments should provide valuable information and thereby benefit the company and its workforce.

Finally, as the use of computers is developing rapidly, it is important to mention computerization in this area. Many COSHH packages are now available of varying price and quality. No ready-made packages can do the assessment for the user so that considerable thought must be applied before contemplating the computerization of COSHH record systems.

The advantages and disadvantages of computerized COSHH systems are summarized in Fig. 9. The biggest disadvantage with most COSHH systems is that of entering information from material safety data sheets into the system. On a continuing basis, this may not present any difficulties but, initially, the task can be a daunting one. Using a computerized system also means that assessments will tend to be carried out in a particular way. This has advantages in that everyone will know where to find the relevant information. However, it can limit the flexibility of the system.

The major advantage of a computerized system is the ease with which one can accurately identify the people who have been exposed, to what substances, at what

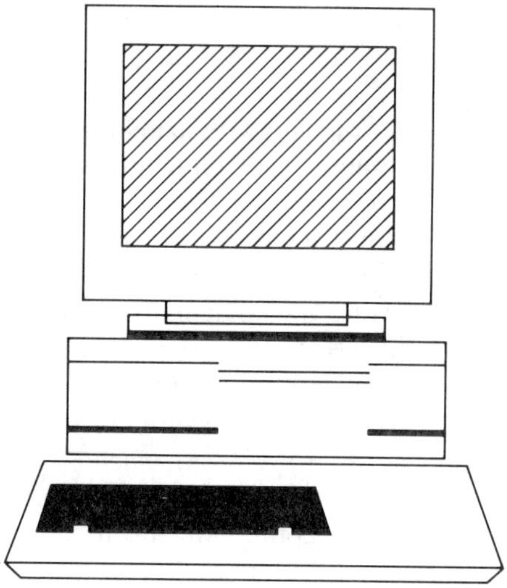

ADVANTAGES

Health monitoring and exposure records can be linked, providing data more easily

Once the system is set up, updating takes less time

Built-in schedulers

Can produce workplace material safety data sheets

Can produce reports

COSHH and standard test data, e.g. on local exhaust ventilation, can be kept together

Standard forms so data easy to locate

DISADVANTAGES

Some data needs to be held twice, depending on the system chosen

Initial set-up may take some time

It is less likely that a computer-based system would be flexible enough to deal with incoming legislation

Usually centralizes information

Subjective judgements are still required

Initial cost is high

Fig. 9. Advantages and disadvantages of a computerized COSHH system.

levels and when. Action points can be automatically re-prompted. As mentioned above, assessments and safety data sheets are produced in a standard format, which means that everyone knows where to find the information. Using a computer also allows reports to be produced easily. Thus, one can readily produce lists of hazardous chemicals which are to be eliminated over a period of time or a list of everyone who has been exposed to levels, say, exceeding 75% of the occupational exposure limit of a particular chemical. A computerized system also simplifies the process of re-assessment as the bulk of the form can easily be copied. Reduced workplace safety data can easily be produced for use at the place of work.

There is no doubt that information can be accessed and manipulated very easily using a computer. However, computerized systems also have their disadvantages. They tend to centralize information when this needs to be available at the point of use. There is also the potential problem that information can remain on someone's desk waiting to be processed. Delays of this sort can encourage people to by-pass the system. The HSE have criticized computer systems as they have found that people were buying systems as a stock answer to COSHH. Programmes can only be a tool in the implementation of COSHH; subjective judgements, and hence training, will still be needed. The introduction of data creates a high initial workload and forces people to use a particular format where a more flexible one may be required to cover different needs.

The final advantages of a paperwork system may be cost, although, over a period of time, this may be debatable. The main advantage of a paperwork system is its flexibility.

When deciding what system to install, each company needs to examine the costs and benefits of each system. One must decide how the system is to be operated and on what it should be run. The most prolific area in COSHH computerization is the stand-alone, personal computer (PC). There are a great many PC solutions available, again at varying costs and quality, the two features not always being linked. The choice of system has to be an individual one. PC-based solutions are likely to be useful only to relatively small, single-site-based companies. Otherwise it is easy for the systems to get 'out of step' and for unnecessary duplication of effort to occur. Another aspect to watch for is that the system should allow area-based assessments and generic assessments to be done. A system which can only handle substance-based assessments will create a great deal of work in anywhere but a small plant. PC-based solutions are usually cheaper than other systems.

The middle-range option is that of networks. Again, with current technology, this is only useful on a one-site basis. However, it can have many users and avoids the problem of getting 'out of step' and duplicating effort. With this solution, costs depend on the number of users.

The final option is that of main-frame systems. Although extremely expensive, these can be used for multi-site companies, spread over a large geographical area. A disadvantage of these systems is that they are less flexible than PC-based systems; any changes to the system parameters require a significant amount of work and can impact on many people. Advantages include the sharing of data, effort and the fact that back-ups are generally handled by the computer department.

10. CONCLUSION

This chapter has provided a brief overview of assessments, highlighting the key principles involved. The key to the process is information; if adequate information can be obtained, the assessment duty is straightforward. Without this, the process of assessment becomes virtually impossible.

APPENDIX

The Control of Substances Hazardous to Health Regulations 1988 introduced two new types of airborne exposure limits; these are:

1. *The Maximum Exposure Limit (MEL)*
   A MEL is the maximum concentration of an airborne substance, averaged over a reference period, to which employees may be exposed by inhalation under any circumstances. The employer also has a duty to take all reasonable precautions and to exercise all due diligence to ensure that exposure is kept as far below the MEL as is reasonably practicable.

2. *An Occupational Exposure Standard (OES)*
   An OES is the concentration of an airborne substance, averaged over a reference period, at which, according to current knowledge, there is no evidence that it is likely to be injurious to employees if they are exposed by inhalation, day after day, to that concentration.

   For a substance which has been assigned an OES, exposure by inhalation should be reduced to that standard. However, if exposure by inhalation exceeds the OES, then control will still be deemed to be adequate, provided that the employer has identified why the OES has been exceeded and is taking appropriate steps to comply with the OES as soon as is reasonably practicable.

REFERENCES

[1] 'Hazard and Risk Explained', HSE Guidance Leaflet, IND(G)67(L), 1988, HSE.
[2] 'COSHH Assessments: A Step-by-Step Guide', HSE Guidance Booklet (ISBN 0 11 885470 4), 1988, HMSO.
[3] 'COSHH in Laboratories', (ISBN 0 85186 3191), 1989, Royal Society of Chemistry.
[4] 'Monitoring Strategies for Toxic Substances', HSE Guidance Note EH 42, (ISBN 0 11 885412 7), 1989, HMSO.
[5] 'COSHH Subject Catalogue', HSE Library and Information Services.
[6] 'Toxicity and Safe Handling of Rubber Chemicals', Third Edition, 1990, British Rubber Manufacturers' Association.
[7] *Sax's Dangerous Properties of Industrial Materials*, Eighth Edition, by R. J. Lewis, 1992, Van Nostrand Reinhold.
[8] 'The Merck Index', 11th Edition, 1989, Harcourt Brace Jovanovich Ltd, London.
[9] 'Hazard Data Sheets', 1991, Merck Ltd, Poole, Dorset.
[10] 'Classification, Packaging and Labelling of Dangerous Substances Regulations 1984', 'Information Approved for the Classification, Packaging and Labelling of Dangerous Substances for Supply and Conveyance by Road,' Third Edition, 1990, HMSO.
[11] 'Control of Substances Hazardous to Health Regulations 1988 and Approved Codes of Practice', 'Control of Substances Hazardous to Health and Control of Carcinogenic Substances', Third Edition, L 5, (ISBN 0 11 885698 7), 1991, HMSO.
[12] 'Occupational Exposure Limits', HSE Guidance Booklet EH 40/92, (ISBN 0 11 885690 0), 1992, HMSO (revised annually).

8

Ventilation systems

Mr J. G. Lyons

SUMMARY

Where substances which are potentially hazardous are being used, it is essential that adequate measures are taken to control employee exposure. The most common method of control is to use ventilation systems. The most popular of these is local exhaust ventilation. These systems need to be properly designed and maintained or they will not provide adequate control and thus employees will be put at risk. This chapter reviews the options available.

1. INTRODUCTION

The assessment of potential risks to employee health, where substances hazardous to health are being used, has been covered in the previous chapter. This assessment should identify the need for engineering controls, such as local exhaust ventilation (LEV). This chapter covers the engineering control options available.

Where substances hazardous to health are being considered for use or actually in use, the first consideration should be to determine whether there is a practical substitute for the material or the process which will reduce the potential hazards. For example, in the use of solvents for degreasing metal components, it may be that a less noxious solvent can be used which is equally effective. Care must be taken, in making this choice, to consider not only toxicity but physical properties, such as boiling point and vapour pressure. The less hazardous solvent may have a significantly higher vapour pressure than the more hazardous one so that its atmospheric concentration would be greater. This could nullify the expected benefits of the change.

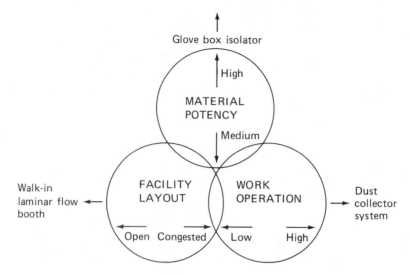

Fig. 1. Containment selection system.

In the handling of solids, substitution of a fine powder by a granular or crystalline solid may reduce the risk of dust exposure to the operator.

Where possible, exposures to all materials should be reduced to levels as low as reasonably practicable and at least below specified limits. Thus, if a process can be automated (so that no-one is involved in any task where contact can occur) or is fully contained (such that exposure cannot occur), there should be no significant risks to health. For example, instead of hand-tipping 25 kg sacks, it may be possible to obtain the material in 1000 kg bags, for which there will be a greatly reduced frequency of risk of exposure. Alternatively, there may be a sufficiently high volume of 25 kg sacks to justify the installation of an automatic 'rip and tip' operation. In this way, human involvement in the task is removed and, with it, the exposure risks.

Total segregation of the employee from the substance should be applied where there is a high hazard associated with the material (see Fig. 1).

However, there are many tasks for which the material/human interface cannot totally be removed and, in these circumstances, the separation of the employee from the hazardous material must be controlled by ventilation systems. There are three main types of ventilation systems which can be used individually or in combination with each other. These are:

— natural ventilation,
— general mechanical ventilation, and
— local exhaust ventilation.

These are discussed below.

2. NATURAL VENTILATION

Natural ventilation, through doors, windows and other leakage in and out of buildings, does not provide a consistent air change rate as for example when comparing summer and winter conditions. Therefore, in general, this approach cannot be regarded as suitable for use as a method of control of exposure to hazardous substances. For example, in a warehouse which has lorry access and internally used, diesel-powered fork lift vehicles, fumes from diesel engines may be retained within the warehouse depending on the direction of the prevailing wind. Natural ventilation is not reliable as a means of providing adequate control of this problem; mechanical ventilation is required.

3. GENERAL MECHANICAL VENTILATION

General mechanical ventilation is used where sources of fumes are diffuse and are relatively innocuous. Under these circumstances, hazardous vapours can be diluted by reasonable air change rates through buildings. However, the design of such systems still needs care to ensure that all relevant parts of buildings are adequately serviced.

Often these systems—especially for large, open-structure buildings—consist of extraction out through the roof only. However, this type of extraction only works effectively where the air rises thermally and/or there are permanent sources of inlet air. Also, this type of system can have economic problems due to heat loss where intake air has to be treated by filtration and/or tempering. Consequently, a balanced system of intake and extraction is the preferred option, the balance of recycled air and fresh intake air, depending on the particular circumstances of the installation.

To reduce the amount of ducting involved, it is becoming quite popular to bring air in by jetting it downwards from the ceiling at the sides of the building and extracting air from the centre of the ceiling. The air is treated and recycled with a suitable bleed and fresh air make-up system. These systems utilize the larger air mixing effects of inlet air, as compared with extraction, to provide the mixing in the work areas. This approach works well, provided the jet units are directed properly; if they are not, dead zones, with no air circulation, can occur (see Fig. 2). Most successful applications of this type use a minimum air change rate of 4 to 6 air changes/hour.

4. LOCAL EXHAUST VENTILATION

The vast majority of potential exposure sources should be controlled at the release point, rather than allowing hazardous materials to escape. Thus, local exhaust ventilation (LEV) is probably the most widely used control measure. LEV should be designed to remove noxious materials from the breathing zone of employees. Extracted air must be replaced with fresh air. Therefore, to protect employees properly, the design of the overall airflow pattern should be such that clean, incoming air reaches the employee's breathing zone before it becomes contaminated by the hazard source. Thus, extraction systems must be designed to draw contaminated air away from the employee. This type of system is probably best illustrated by the rear extract laminar flow booth, discussed below.

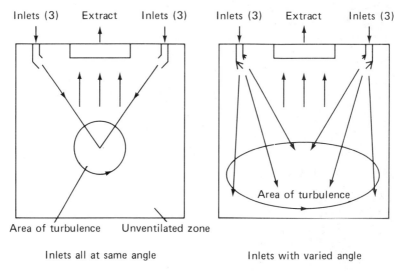

Fig. 2. General ventilation showing need for directional control of inlets.

4.1 Laminar flow booths

In this type of booth, the entire rear panel forms the extraction zone giving a uniform flow towards the rear of the booth. Provided the employee stands so that the air flows from one side to the other, he will be protected. However, he can get into an unprotected position between the task and the rear extract. In addition, by standing in line (upstream), he can also be less well protected due to distortion of the airflow pattern by his body which can cause a low pressure zone to be set up in front of him. This can cause dust or fume from the task to rise into his breathing zone. Therefore, where rear extract laminar flow booths are used, the task layout should be arranged so that the employee cannot get into a dangerous position (i.e. the shaded positions in Fig. 3).

In addition, the source of incoming air must not only be sufficient but also not distort the laminar flow towards the rear of the booth. In the example in Fig. 4, the rear extract laminar flow booth is located at right angles to the supply air thus causing a swirling action within the booth itself. This reduces its effectiveness.

4.2 Downflow booths

A modern approach to this problem is to use downflow laminar booths. In these, the clean intake air comes as a uniform downflow from the ceiling of the booth and the air is extracted at floor level from the rear (see Fig. 5). This configuration makes sure that the employee's breathing zone is always on the clean side of the task and that any dust generated, which will settle anyway, is carried downwards away from the breathing zone. This approach also makes more effective use of intake air to move particulates away from employees than extract alone. Commercially available units of this type have either a single pass air system or a full treatment and recycle air system. The treatment systems involve at least two-stage treatment with a main filter unit backed up by a high efficiency filter (see Fig. 6).

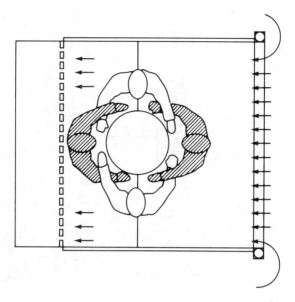

Fig. 3. Horizontal laminar flow booth (safe operation positions at the sides only, unshaded).

4.3 Specific extract systems

There are many work tasks where the booth approach is not a suitable solution so that extract ventilation needs to be designed to fit the task. The design of such extract systems needs to be done with care; there are many examples which provide inadequate control due to poor design.

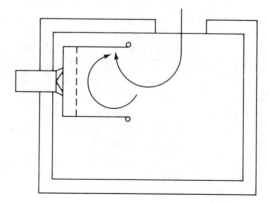

Fig. 4. Laminar flow booth performance is unpredictable in confined rooms.

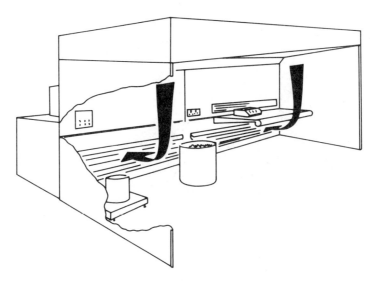

Fig. 5. Laminar downflow booth.

The use of canopy hoods over tasks which do not involve thermal rise is a common mistake. In such situations, most of the extracted air movement occurs around the edge of the canopy hood and thus does not control the escape of materials which can easily get into the operator's breathing zone.

The use of a small canopy hood over a conveyor interchange has little effect on dust escaping from around descending powder. The installation of side and rear panels and a reduction in the front opening make this into an effective control (see Fig. 7). The use of a canopy hood over a degreasing tank potentially puts the operator's breathing zone in between the source and the extraction point. This configuration

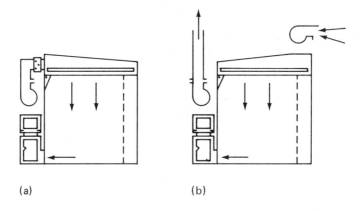

Fig. 6. Downflow systems: (a) treatment and recycle; (b) single pass system.

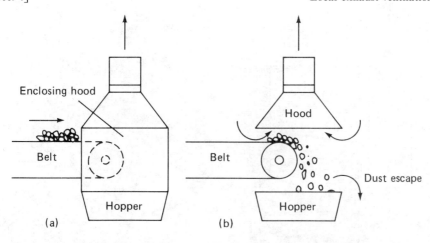

Fig. 7. Conveyor transfer point: (a) effective control with enclosure; (b) ineffective canopy hood.

gives the employee an undesirable exposure to the vapour. A rear extract slot (or, if the tank is wide, an inlet and extraction system across the top of the tank), can give adequate control and thereby reduce employee risk of exposure (see Fig. 8).

It is also important to consider all aspects of the work task and provide adequate control for the whole operation.

The emptying of 25 kg sacks into a hopper is a very common operation in many factories and there are many elegant ventilated tip hopper units. However, they often do not provide a safe disposal method for the empty bag. Consequently, the operator may resort to removing the air from the bag and folding it up etc., creating a dust cloud in and around the area for both the operator and other employees to breathe.

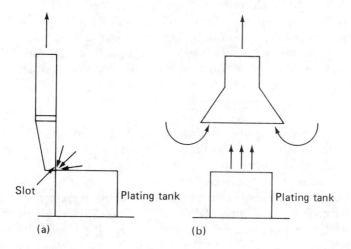

Fig. 8. Solvent tank ventilation: (a) effective rear slot control system; (b) ineffective canopy hood.

134 Ventilation systems

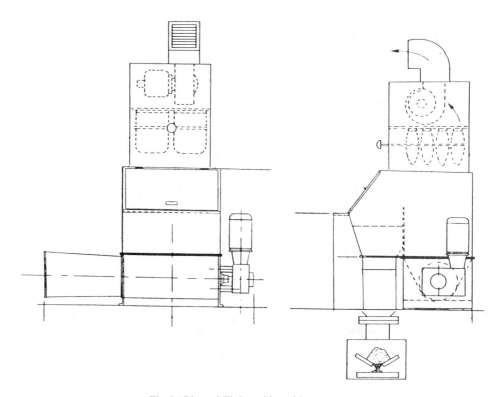

Fig. 9. 'Rip and Tip' machine with compactor.

In addition, the dust will settle on the floor and other surfaces from where it will eventually have to be cleaned up. For little extra cost, at the time of purchase, a ventilated tip hopper unit, complete with a waste compaction unit (all inside the ventilated control area), can solve the problem (see Fig. 9).

There are proven designs for many tasks; a large range, with full operational specifications, is given in 'Industrial Ventilation', published by the ACGIH [1].

4.4 Portable extraction systems

There are many industrial situations where work tasks are of sufficiently short duration that portable extraction equipment can be a sound choice. The requirements of airflow (discussed above) still apply. In addition, care must also be taken to ensure either that the extracted air is removed safely from the work area or that the treatment system is suitable for removing contaminants prior to recycling.

Portable units for the control of dust, which use a cartridge filter prior to recycling to the working environment, do not remove gases. They should also be fitted with a secondary filter to ensure that any dust escaping through the primary filter does not re-enter the working environment.

4.5 Laboratory fume cupboards

A specialized, local exhaust ventilation system is the laboratory fume cupboard, the installation and use of which need special care. The basic design and installation criteria are covered by a British Standard [2] which gives details of the smooth shape and air velocity requirements required to give the requisite performance. It is important that, in the fitting of additional equipment around a fume cupboard, due consideration is given to its effect on the airflow characteristics of the unit. The shaped panels at the sides are designed to give a smooth airflow. This can be seriously impaired by the mounting of power sockets on these panels. This can create a turbulent air zone which reduces the efficiency of the overall performance.

Fume cupboards can take a considerable volume of air from the room (typically 0.4 m^3/s) so that make-up air is usually required. The positioning of inlets into the room is critical. Air should be brought into the room in such a way that it does not affect the uniformity of airflow extracted at the fume cupboard sash opening. For installations, especially those having more than one fume cupboard to a room, this is often achieved using the ceiling void as a plenum chamber with a perforated plate ceiling for the air intake. Once installed, fume cupboards often take a considerable time to achieve the correct balance between the top and rear extract points. In addition, 'bagging out' of some perforated plate ceiling panels may be required to avoid downflow turbulence affecting the performance of the fume cupboards.

For energy saving purposes, and to avoid high air velocities when the sash is closed (or nearly closed), many installations are now being fitted with variable flow systems. These systems offer advantages in that reduced air volumes are extracted; lower air velocities in the sash-closed position reduce turbulence problems around the bottom sill aerofoil. However, their speed of response to the opening of the sash, if too slow, can allow contaminated air to be released into the room. Fume cupboards are designed as an emergency release or spillage control measure and should be assessed for their performance under these conditions.

5. TREATMENT OF EXHAUSTS

Ventilation systems, especially those which recycle air to the working environment, require systems to treat the exhaust air depending on the material being controlled. Solids/dusts are often treated by dry methods such as bag or cartridge filters. These filters often have secondary 'back up', high efficiency filters especially when the air is being recycled. Gases and vapours are often treated by wet methods which can be of several types [1]. In some cases, such as liquid aerosols, electrostatic precipitators may be used. This approach is also often used in the treatment of welding fumes. In this application, solid particles are removed efficiently but gaseous emissions from the process are not treated. Therefore, if the particular welding process gives off noxious gases, the treated air should not be recycled to the working environment.

It is clear that the selection of air treatment methods must be chosen with care and located carefully. For example, the type of dust collection method required for a dry organic dust should be designed to minimize potential ignition sources (e.g., earthed to avoid the generation of static electricity) and should have an explosion vent located suitably to release to a safe area, usually outside the building, etc. Standard system designs and their applications are provided in Reference 1.

After treatment, extracted air is often discharged directly out of the building. Careful consideration of the point and velocity of such discharges is critical especially in relation to intakes into the building. This can be examined by the use of dispersion models but, sometimes, other techniques such as wind tunnel testing, may have to be utilized.

6. MAINTENANCE OF VENTILATION SYSTEMS

Ventilation systems must give a clear indication to users of the correct operation; air velocity is often the parameter of choice. However, the instrumentation required is susceptible to damage and failure. The alternative, for which there is robust and reliable instrumentation, is to measure static pressure. Therefore, local exhaust ventilation systems are now routinely fitted with a static pressure gauge marked with an acceptable performance zone. In this way, the operator has a clear indication of satisfactory performance. The static pressure display not only gives the operator confidence that the device is operating properly but will clearly indicate a malfunction, for example, a belt slippage between the motor and the fan. Once a ventilation system has been properly commissioned, the static pressure gauge can also be used as a measurement of continued good performance for routine checks as required by COSHH, etc.

Ventilation systems, like all process plant, require routine maintenance. If the ventilation system is a control measure for a hazardous substance, there is a legal requirement to examine and test the system thoroughly at least every 14 months. Guidance on the requirements for this work is given in HSE guidance notes [3].

REFERENCES

[1] 'Industrial Ventilation' 20th Edition, the American Conference of Governmental Industrial Hygienists, 1988.
[2] British Standard 7258 on Laboratory Fume Cupboards, Parts I, II and III, 1990, BSI.
[3] 'The Maintenance, Examination and Testing of Local Exhaust Ventilation', HSE Guidance Booklet HS(G) 54 (ISBN 0 11 885438 0) 1990, HMSO.

ACKNOWLEDGEMENTS

Figures 1, 3 and 4 are reproduced by kind permission of Extract Technology Ltd. Figures 5 and 6 are reproduced by kind permission of Vitalair Ltd.

9

Personal protective equipment

Mr P. J. Turnbull

SUMMARY

The use of personal protective equipment (PPE) to provide protection against hazardous substances should always be regarded as a 'last resort' when other control measures, such as engineering controls or systems of work, are inadequate or not reasonably practicable. Consequently, a poor PPE programme may result in worker exposure to hazardous substances, possibly with fatal consequences.

Development of a successful PPE programme requires consideration of a variety of key elements, including hazard and risk assessment, PPE selection, purchase and supply, training and wearer appreciation and maintenance and disposal. Correct selection of PPE requires a thorough understanding of the hazards and tasks for which the PPE will be provided, the potential performance of the PPE and any statutory requirements or codes which need to be observed. The needs of the wearer must be considered thoroughly, as uncomfortable or inconvenient PPE will be worn incorrectly or not at all, defeating its purpose.

Compliance with statutory requirements, including European Community directives and standards, has a direct effect on the selection of PPE. The development of PPE specifications for European approvals is explained.

The final, and most important, part of any programme involves the routine use of PPE in the workplace. Wearer training and hazard appreciation are vital to the successful use of PPE and suitable support systems for purchase, supply, storage, maintenance and disposal are necessary to complete an effective PPE programme.

1. INTRODUCTION

The use of personal protective equipment (PPE) to provide protection against hazardous substances should always be regarded as a 'last resort' when other control measures, such as engineering controls or systems of work, are inadequate or not reasonably practicable. The principle that PPE should always be regarded as the last in the line of defence when protecting people against exposure to hazardous chemicals is often quoted but, unfortunately, not often followed. All too often the first reaction is to ask 'Which gloves should we use?' or 'Which respirator is best?' rather than 'Can we do this job without being exposed to contaminated objects?'. Often, the quickest and cheapest solution to a chemical exposure problem is to consider PPE as the first, and possibly only, option.

Current UK and European safety legislation places great emphasis on a hierarchy of measures to prevent or control exposure to hazardous substances. The Control of Substances Hazardous to Health Regulations 1988 (COSHH) [1] presents a list of methods for controlling exposure and the use of PPE is relegated to tenth out of twelve possible methods. However, PPE may be the only reasonable option for many circumstances, e.g., emergencies, etc., where the correct type of PPE and the way it is worn are vital, as exposure to the hazardous chemical may have serious or even fatal consequences. Even under mild conditions of use, people rely on the effectiveness of PPE and faulty or badly worn PPE may result in low level, long-term chemical exposure.

While recognising that PPE should be low on the list of ways to control workplace exposure to hazardous chemicals, there are still many legitimate reasons for needing to provide PPE to the workforce. The development of an effective PPE programme is vital to ensure that:

(a) the need for PPE is clearly defined,
(b) the correct PPE for the task is selected,
(c) the wearer is fully trained for the PPE to be used, and the reasons for needing PPE, and
(d) a suitable supply, maintenance and disposal procedure is defined to support the PPE.

Failure in any of one of these will result in an ineffective PPE programme and may result in the wearer being exposed to hazardous chemicals.

2. PERSONAL PROTECTIVE EQUIPMENT

Of the three possible ways in which hazardous chemicals can enter the human body, personal protective equipment (PPE) can reduce or prevent exposure through inhalation and absorption. In many circumstances, the chemical may be absorbed through the skin and may also present a respiratory hazard. Any PPE selected will have to protect against the combined hazard, either as a single item of PPE, or as a combination of two or more types. Very rarely are we faced with a single, clearly indentifiable hazard and it is therefore important that the assessment of a task prior to PPE selection considers the possibility of multiple chemical hazards, either from the task being assessed or from nearby work.

Very often people in industrial environments will also be required to wear PPE to protect against non-chemical hazards, e.g., noise, impacts or falls. It is important to consider the compatibility of all items of PPE to ensure that interference between different items does not reduce the protection afforded by any single item.

The provision of PPE in the workplace may mean that, perhaps, several items of PPE may need to be matched to provide total protection. Consequently, to avoid confusion, respiratory protection and skin protection are considered separately in this chapter.

3. ASSESSMENT

It is vitally important, when carrying out an assessment for the use of PPE, to remember that the assessment is of a task at a particular moment. If, for any reason, that task, or any of the immediate surroundings, is altered, the assessment will be invalid. Furthermore, routine reviews of tasks should be carried out and assessments and PPE requirements amended, if necessary.

It is necessary, before proceeding with any PPE selection, to ensure that a clear understanding of the terms chemical 'hazard' and 'risk' is maintained.

HAZARD: The potential of a chemical, derived from its intrinsic properties, to cause harm to the human body.

RISK: The probability of a specific, undesired event occurring so that a chemical hazard will be realized (so as to cause harm to the unprotected person) during a stated period of time or in specified circumstances.

For example, a concentrated acid presents a high chemical hazard because its potential damage to the human body (if spilt on unprotected skin) is very severe. If, however, the acid is contained in an acid-resistant, impact-proof sealed container, the probability of the acid being spilt is low; consequently, the risk is low. To include other factors in the overall picture, e.g., quantity of chemical spilt, or how the chemical is spilt (e.g., a jet or spray), the term 'danger' is often used. This is a vague term and, although often misused, is suitably descriptive.

It is important, when carrying out an assessment, to ensure that the real consequences of the spill or exposure are clearly defined and properly recorded. Often, people only look at the material safety data sheet and forget that their process stream may be at elevated temperatures or pressures which will clearly alter the real consequences of a spill or gas escape.

3.1 Assessment of chemical hazards

A wide variety of sources of chemical hazard data are available, especially since the coming into force of the COSHH Regulations and the amendments to Section 6 of the Health and Safety at Work etc. Act 1974 [2]. Although these basic data are useful, it is important to ensure that the assessment takes account of the *actual* condition of the chemical, e.g., in the process stream and when it is released into the surrounding area!

It is important to identify how the chemical will affect the human body. For example, the speed at which the chemical may have an effect on the body will be important in deciding whether the chemical protective clothing can allow some leakage or none at all. Although it may seem advantageous to have no inward leakage, totally liquid-tight garments can be very hot and uncomfortable. To require such suits to be worn continuously, in case there may be an emergency, is difficult to enforce. It may be more realistic to tolerate a garment with lower protection if the chemical does not have an immediate effect on the skin and can be washed off without long-term health effects.

With inhalation hazards, data on occupational exposure limits is often used to determine the degree of protection necessary from respiratory protective equipment using the concept of protection factors. This information may be gathered from HSE-published data [3].

3.2 Assessment of risks and danger

It is important that all practicable measures to reduce the probability of the chemical hazard coming into contact with the human body are explored before considering the use of PPE.

Assuming that all practicable measures have been taken to reduce the risks, it is important to consider the likely duration and frequency of exposure. For instance, under routine working conditions, chemical protective gloves may be splashed at regular intervals during the working shift because of a particular task. It may be possible to select a heavy type of glove which will withstand repeated splashing for a few hours or there may be a less protective lightweight glove which could be replaced at regular intervals. Either type may be acceptable but other features, such as dexterity and comfort, may be critical.

Similarly, one needs to consider if an emergency occurs, whether personnel will be expected to carry out emergency shut-down procedures *before* escaping or whether escape will be the primary objective.

It can be seen, from the examples quoted above, that carrying out the assessment is critical, as there are many factors in the selection of PPE which require balancing to achieve the best solution. Unfortunately, there are no hard and fast rules or sequences which can be followed to generate the perfect assessment. British Standard 7184:1989—Selection, Use and Maintenance of Chemical Protective Clothing [4], gives a useful flowchart which provides the basic questions needed to complete an assessment for chemical protective clothing, which can be generalized to include all PPE.

4. STANDARDIZATION AND LEGISLATION

Requirements to provide PPE are given in various UK and European safety laws. Without quoting the detailed requirements of the COSHH Regulations or the European directive on the Use of PPE [5] (and the corresponding UK implementing regulations), common themes run through them.

They all require that PPE should be selected, after appropriate assessments have been made, to ensure that it is suitable for the purpose. Furthermore, they require that PPE is approved to a performance 'standard', in most cases a common European standard.

To understand fully how and why European standardization has taken place one has to reflect on the history of standardization and the political decisions made by the European Commission. Historically, the standardization of PPE outside the UK was split between the International Standards Organization (ISO) and the Comité Européen de Normalisation (CEN). Until April 1989, standardization of protective clothing was solely the responsibility of ISO, while CEN concentrated its work on respiratory protective equipment. As part of the development of the Single European Market, the Council of the European Communities (EEC) recognized that comparable standards of quality and performance of many commodities with were needed throughout the whole of the Community. In the area of personal protection, the Council adopted two directives, one concerned with the manufacture of PPE (commonly called the 'Article 100' directive [6]), and the other with its use (commonly called the 'Article 118A' directive). A fundamental feature of these directives is the provision of approved PPE. Essentially, manufacturers of PPE cannot sell equipment which has not gained a 'CE' mark of approval, which shows that the equipment has satisfied the requirements of a testing specification, and users cannot use equipment which has not obtained this mark.

4.1 Respiratory protective equipment
The most advanced programme of standardization can be found with respiratory protective equipment (RPE). The CEN committee responsible (TC 79) has been developing standards for almost 20 years and was therefore well prepared when the new EC directives were formulated.

The major problems facing users of RPE are:

(a) the variety of equipment available and the attendant confusion which can arise without thorough training and education, and
(b) the misconception that RPE, which is approved to a standard, is suitable for use.

This final point concerns the misconception that equipment which meets the appropriate European standards is suitable for use. The primary aim of standards technical committees is to develop specifications which will provide a consistent benchmark for RPE, which may be tested in laboratories anywhere throughout the European Community. Inevitably, they set reasonable minimum standards to ensure that ineffective and dangerous equipment is eliminated from the marketplace. However, when it comes to specifying suitable protective equipment, only the users can assess equipment performance by in-house testing and field trials.

4.2 Chemical protective clothing (CPC)
The standardization of protective clothing in Europe has seen the most dramatic activity in recent years. Prior to April 1989, there were no standards for protective clothing but, by November of that year, 23 documents had been approved by CEN

Technical Committee TC 162 for circulation as draft European standards. By mid-1991, there were 67 standards at various stages of preparation covering protection of the body.

Draft European standards for CPC have been developed in a different way from most other European standards. In an attempt to accommodate a wide range of CPC, several distinct types of protective clothing were identified.

To assist users, performance levels have been set above a basic minimum level for many of the important testing criteria. Manufacturers have their CPC tested using standard test procedures and performance levels reported for each test. The test results can then be presented in the form of a table which allows users to select the most appropriate CPC for their purpose.

4.3 Combined PPE
One of the most difficult areas of CPC standardization, to date, has been the development of standards for equipment which provides both body and respiratory protection. In the chemical industry, the presence of one or more hazards is very common and yet standardization of PPE has taken place in at least five technical committees. For example, it is interesting, if not dismaying, that none of the respiratory standards for airline-fed respiratory protection includes any performance testing against chemicals.

5. RESPIRATORY PROTECTIVE EQUIPMENT

If it is found that chemicals present in the assessed task present a respiratory hazard, which cannot be eliminated or reduced to an acceptable level by other control measures, it will be necessary to consider the use of respiratory protective equipment (RPE). This section identifies the major types of RPE available and summarizes their advantages and disadvantages.

Dusts, fumes, vapours and gases encountered at work can easily enter the body by inhalation. A very wide range of RPE has been developed to deal with situations ranging from nuisance dusts, in low concentrations, through to lethal concentrations of acutely poisonous gases. Consequently selection of the correct RPE for the task is often difficult and misunderstood.

5.1 Types of respiratory protective equipment
RPE can be divided into the following two major groups according to how it protects the wearer (see Fig. 1):

(a) those which filter hazardous substances from the air, commonly called respirators and
(b) equipment which supplies clean, uncontaminated air to the wearer through flexible pipes or from cylinders.

These groups can then be divided to accommodate the wide variety of RPE which has been developed to satisfy the range of requirements.

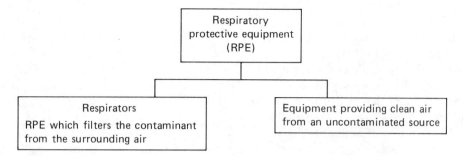

Fig. 1. Types of respiratory protective equipment available.

5.1.1 Respirators

Respirators only remove contaminants from the surrounding air. They do not supply oxygen, so they cannot be used in oxygen-deficient atmospheres. (This must never be forgotten.) Respirators can be divided into two groups; those which rely on the power of the wearer's lungs to draw air through the filter and those which use a small motor or blower to force air through the filter(s). The first group are often called simple filtering respirators and the second group, powered respirators (see Fig. 2).

Further sub-divisions below this can be made depending on how the filtered air is fed to the wearer's lungs or whether the device is disposable.

How respirators work

Respirator filters are available in a variety of types. Gas filters remove gases and vapours; particle filters remove dusts and other particulates including aerosols. Combined filters remove particles, gases and vapours.

Particle filters work by trapping particles on to the filter material by a variety of mechanisms, including mechanical impingement and/or electrostatic charges. Some filters are only suitable for solid- and water-based aerosols and should be marked, 'For use against solid aerosols only'. The efficiency of a particle filter can be measured

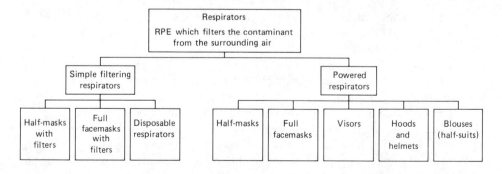

Fig. 2. Types of respirators. Note: particle filters are available in three grades of filtering efficiency. Gas filters are available in three grades of capacity (size).

by standard laboratory tests and the filter classified according to its laboratory performance. Particle filters to the European Standard BS EN 143 [7] are classified at three levels; low, medium and high efficiency, according to how much of the laboratory test challenge is retained by the filter. Although it may seem logical to use the highest efficiency filter possible to ensure maximum protection, it should be borne in mind that higher efficiency filters generally have high breathing resistance which may be less comfortable to the wearer.

The filtration efficiency of the filter depends on the particle size of the contaminant. Very small particles, such as fumes of condensed vapours from welding or molten metals, will penetrate the filter more easily than standard laboratory test aerosols. A further problem with particle filters is that filter 'clogging' may increase the breathing resistance as particles become trapped on the filter. Some filters are specially designed and tested to resist clogging.

Gas and vapour respirator filters
Gases and vapours can be removed by gas respirator filters in the following principal ways:

(a) **Adsorption**, the most common method, predominantly retains contaminant molecules on exposed surfaces of the sorbent granules by physical (physi-sorption) or chemical (chemi-sorption) attraction. With physi-sorption, the strength of the physical bond depends on the type of sorbent (usually activated carbon) and the contaminant. The bond may easily be broken, releasing the hazardous gas and vapour molecules. To increase the strength of the bond between the contaminant and the sorbent, the sorbent may be impregnated with a chemical which will form a stronger chemical bond (chemi-sorption). This type of mechanism can be used to produce sorbents which are more selective of the contaminants they can trap.
(b) **Absorbents** differ from adsorbents in that, although they are porous they have a smaller surface area. As the name implies, gas or vapour molecules penetrate deeply into the molecular spaces throughout the sorbent and are held there chemically. Most absorbents are used for acidic gases and include mixtures of sodium or potassium hydroxide with lime and/or caustic silicates.

Filters are available for a wide range of uses and are defined in European Standard BS EN 141 [8]. Although filters are classified as, e.g., organic filters, acid gas filters, etc., it is important to determine that the filter selected is appropriate for the specific chemical duty. This information should be provided by the filter manufacturer although some in-house testing should be undertaken to prove its 'suitability for use' under workplace conditions.

It is a widespread view that gas respirator filters are 100% efficient until the sorbent's capacity to absorb the gas and vapour are exhausted. Then, contaminants will pass completely through the filter and into the face-piece. However, as a large majority of gas respirators use filters which employ physi-sorption as the principal mechanism, and it is this mechanism which is the most easily reversible, the retention of the toxic gas or vapour on the filter cannot always be assured. Any factor which may affect the physical bond of the contaminant to the sorbent must be recognised as a threat to filter efficiency. Increased humidity, unusual temperature conditions and the presence of other contaminants can seriously affect filter efficiency and capacity.

Respirator types

It can be seen from Fig. 2 that the 'family' of respirators contains a large number of variants, especially when one considers that, for every type of respirator, there is a wide range of filters (gas, particulate or combined filters) which can be fitted to the respirator. The wide variation of respirators has occurred because the filter respirator is very convenient. However, it also has some drawbacks.

The advantages are: they are lightweight, comparatively simple to use, do not require sophisticated air handling equipment and are inexpensive. The range of filters available is very good and they are simple to replace.

The disadvantages are: they cannot be used in oxygen-deficient atmospheres as they only filter the air; filters become exhausted after some time in use; gas filters fail to danger. Simple filter respirators rely on lung power to draw air through the filter; this can be tiring and requires the wearing of a tight-fitting face-piece. Simple respirators are particularly prone to face-piece leakage whereby contaminated air is drawn into the face-piece through small leaks between the wearer's face and the facemask. Anything which interferes with the facial seal, i.e. spectacles, beards, long hair or unusual face shapes, can result in face-piece leakage and, hence, inhalation of hazardous vapours.

To overcome some of the problems of negative pressure respirators and improve the comfort of the wearer, a range of powered respirators has been developed. Early powered respirators had a simple, battery-powered, continuous-flow fan which pushed dusty air through a filter bag into a loose-fitting helmet and visor. By providing a high flow of filtered air into the breathing zone and thus providing a positive pressure, it was possible to use more comfortable loose-fitting visors, hoods and blouses rather than rubber face-pieces. As these powered respirators used high continuous air flows, they were not practicable for use with gas filters because the high flow-rates quickly exhausted the filters.

More recent developments in powered respirators have led to fan units which feed air only when the wearer inhales and are more economical in filter use. This 'demand' or 'breath-responsive' type will only work with a full facemask but has an advantage over those fitted with hoods, helmets, etc., because the wearer can suck air through the filter(s) if the fan unit stops working. A powered respirator fitted with a loose-fitting hood, helmet, etc., will fail to danger if the fan unit stops working, because the user will suck contaminated air directly into the hood or helmet, by-passing the filters.

5.1.2 RPE providing air from an independent source

This type of equipment supplies breathable air to the wearer from an independent source; from compressed air cylinders, through a flexible pipeline (airline) from an air compressor or by a combination of both methods. As with respirators, the variety of devices available in this category is extremely great (see Fig. 3). The performance of each device depends on the type of facial covering, facemask, hood, visor, etc., and the way the air is supplied to the wearer.

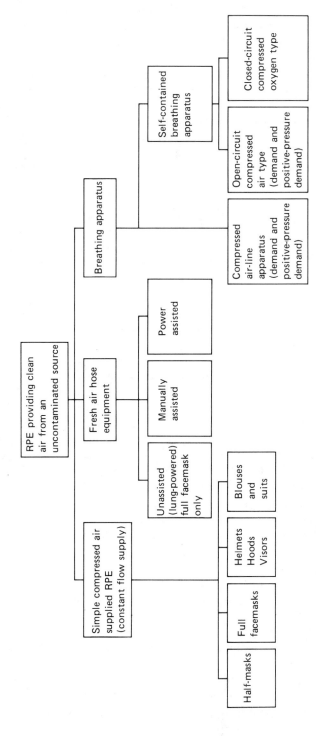

Fig. 3. Types of RPE providing air from an independent source.

Self-contained breathing apparatus
Self-contained breathing apparatus (SCBA) provides the highest level of respiratory protection and is intended for use in the most hazardous situations, including those which are immediately fatal to the unprotected. It is used extensively by fire services and other emergency units.

Open-circuit SCBA supplies compressed air from a high-pressure cylinder carried on a harness and may be supplied to the wearer by:

(a) Negative-pressure demand valve, where suction caused by inhaling opens the demand valve which then floods the facepiece with air. This type necessarily generates a negative pressure in the facepiece at some time during the breathing cycle.
(b) Positive-pressure demand valve, where the mask is maintained at an internal pressure slightly higher than atmospheric pressure throughout the breathing cycle. The demand valve operates as above and still opens on 'demand' but this type ensures a positive pressure under all but the highest work rates.

Closed-circuit SCBA operates by re-circulating the breathing air after absorbing the exhaled carbon dioxide and adding more oxygen. Escape SCBA is available in either open- or closed-circuit types and is intended for escape only. It must never be used for working or to re-enter a hazardous area.

Compressed-air line equipment
The simplest type of compressed-air line device comprises a source of breathing quality air (see section 5.4 below) fed to wearer through a flexible tube (or airline), with a simple flow regulator (either fixed or adjustable) mounted on a harness or waist-belt, and thence to the face of the wearer. The air may be directed into the breathing zone by the use of a blouse, hood, helmet, visor or facemask. More complex, compressed-air line equipment may be fitted with demand or positive pressure demand valves and emergency escape cylinders. This type of equipment offers comparable protection to self-contained breathing apparatus but with longer effective work duration.

The most obvious disadvantages of an airline-supplied system are the wearer's reliance on the air supply system, the source of clean air and the problems of working while trailing an umbilical supply line. One less obvious problem with airline-supplied RPE is the possibility of chemical degradation or permeation of the airline. Hazardous chemicals can become absorbed into the plastic airline and then diffuse into the breathing air supply within the tube.

Fresh air hose equipment
Fresh air hose equipment draws clean air from a source near the contaminated area and delivers it to a facemask or, if a powered fresh air system, to a hood.

5.2 RPE selection
It can be seen, from the wide selection of RPE outlined in Figs. 1–3, that several types of equipment may be suitable for a particular set of conditions. The level of

protection afforded by several different devices may be similar but other factors, such as wearer comfort, work duration, ease of maintenance, economics, etc., must also be considered. There is no simple formula for the selection of RPE—or any PPE for that matter—and only careful balancing of each factor will ensure that the correct equipment is provided.

To identify the basic types of RPE which may be used, a logical sequence of questions can be developed as shown in Fig. 4. Unfortunately, the logic diagram becomes more complex than that shown in Fig. 4 when factors such as filter selection, comfort, durability, maintenance requirements, etc., are included. Balancing all the factors needed for successful RPE selection is difficult and requires an intimate understanding of each type of equipment, its potential protective capability and the supporting systems which need to be established to ensure that the selected devices remain effective.

Guides to RPE selection have been published which contain useful descriptions of types of RPE and their merits [9,10]. Fundamental to all RPE, however, is the use of the term 'protection factor' to describe the potential performance of RPE.

5.3 Protection factors

The performance of respiratory protective devices can be expressed as the ratio of the concentration of the contaminant in the surrounding air (C_o) to that inside the device (C_i):

$$\text{Protection Factor} = \frac{C_o}{C_i}$$

The presence of contaminant inside the mask will depend on various factors such as filter penetration, exhalation valve leakage, other manufacturing defects and leakage caused by the fit of the RPE to the face—face-seal leakage.

There is a great deal of uncertainty as to whether protection-factors provide a reliable indication of RPE performance so that several types of protection factors have been developed [11]. Most protection factors quoted are determined under standard laboratory conditions and not in the workplace where other influences can severely increase the face-piece leakage. The following points should be borne in mind when using quoted protection factors for respirator selection:

(a) the protection factor quoted may not be realized in use and
(b) anything which can interfere with the seal between the face and the RPE, e.g. facial hair, spectacles, etc., will cause leakage of contaminant into the mask.

The fit of the face piece is most critical with simple respirators as they rely on suction inside the facemask to draw air through the filter(s). In the final part of the selection process, fitting the respirator to the wearer, it is important to remember that one facemask will not be suitable for all facial shapes and sizes and that different sizes or makes may be needed to achieve effective protection. Testing for face-piece fit in the selection process is mandatory in some countries and can either be a simple qualitative test, i.e., smell or taste, or a more sophisticated quantitative test which gives a certificate of fit.

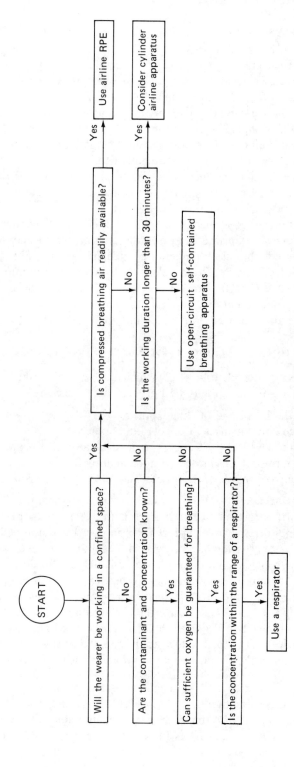

Fig. 4. A typical logic sequence for RPE selection.

5.4 Maintenance of RPE

Apart from disposable respirators, all types of RPE must be maintained to ensure that the original protective capability is preserved. There is a legislative requirement for maintenance to be carried out at regular intervals by competent persons and for records to be kept [1,2,5]. All manufacturers of RPE must provide advice on maintenance and it is important that this information is requested. If maintenance is to be carried out in-house, someone must be nominated and trained to perform this duty. External maintenance services should be assessed, preferably by the RPE manufacturer, to ensure that they are competent to maintain the RPE. Daily visual inspection of RPE is often assigned to RPE users, so that suitable training must be carried out. This must be supported by regular, thorough examinations and maintenance.

Disposal of redundant equipment or exhausted filters should be carried out through a special procedure, appropriate to any contamination which may be present. Any redundant equipment should be rendered completely unserviceable before disposal.

When selecting RPE which provides clean air from an uncontaminated source (e.g. airlines and SCBA), it is particularly important to ensure that the air supplied to the wearer remains of an acceptable quality. Routine maintenance of air compressors and testing of the air quality are important and are also legislative requirements [1].

6. CHEMICAL PROTECTIVE CLOTHING

Five key factors have been identified [12] which are essential for the long-term success of a chemical protective clothing (CPC) programme. These are:

(a) determining the likelihood of skin exposure,
(b) identifying the consequences of direct skin contact,
(c) establishing the levels of protection provided by available protective clothing,
(d) making an appropriate selection and documenting the basis for the selection, and
(e) training employees in the proper use of the selected item or items.

Having established the need to provide some form of personal protective equipment (PPE), it is important to remember that each item of PPE must be compatible with other devices. Only rarely is one type of PPE needed. Invariably, CPC will need to be worn with gloves, boots and face protection. Respiratory protection may also need to be worn; some protective clothing incorporates RPE to ensure compatibility. As a general rule, if there is a respiratory hazard as well as a skin hazard, RPE should be selected before CPC as respiratory hazards are generally more critical. It is vital to ensure that the CPC does not reduce the effectiveness of the RPE.

In this discussion on CPC, although gloves and boots are not considered separately, the general principles described below apply equally to this equipment. However, chemical protection of the eyes is a separate topic and is not included in this section [13,14].

6.1 Permeation and penetration

Chemicals can pass through to the interior of CPC and gloves in two principal ways. Firstly, this can occur through holes in the clothing caused by poor manufacture, physical damage or, simply, through the necessary openings for the head, hands and feet. This leakage is called penetration and can generally be overcome by good design and manufacture. Secondly, it can occur by diffusion through the air-impermeable material of the CPC. This process is called permeation and proceeds in the following three stages:

(a) sorption of the molecules of the chemical into the contacted, outside surface of the material of the protective clothing,
(b) diffusion of the sorbed molecules through the material and
(c) desorption of the molecules from the inner surface of the material.

The time that the chemical takes to pass from the outer surface of the CPC or glove material to the inner surface is called the breakthrough time and the rate at which the chemical passes from one side to the other is called the permeation rate. Breakthrough time values are often quoted as a measure of performance for CPC and gloves. As with respiratory protection factors, there is a great difference between laboratory-generated data and that found in the workplace. Laboratory tests which are used to determine breakthrough time entail keeping the chemical under test in permanent contact with the protective clothing or glove throughout the test. This rarely happens in real life as CPC is generally splashed only intermittently and much of the contaminant evaporates from the outer surface. It is important to recognise that quoted breakthrough times and permeation rates are laboratory values determined under specific testing conditions.

As permeation and penetration reduce the protection afforded by CPC, it is important to include both these factors when considering the potential effectiveness of various types of clothing. After all, it makes no sense to use a garment made of material with a high resistance to permeation if it allows rapid penetration of the chemical to the inside of the suit through poor design features or poor manufacturing quality.

The range of CPC available today grew principally from needs in industries which manufacture or use chemicals and generally evolved as the requirements of industry became more stringent. Developments in the range of protective clothing materials and better understanding of the way that protective clothing provides protection have led to more sophisticated and often ingenious devices offering very high levels of protection.

6.2 Protective clothing materials

Materials used in the manufacture of CPC include textiles, rubber and plastic films and coatings and may be permeable or impermeable to air. Air-permeable materials (e.g., woven and/or spun-bonded fabrics, often including a microporous film), are permeable to gases and may also allow liquids to penetrate, especially if the liquid is applied with force, i.e. a jet. They are, however, more comfortable to wear as they are 'breathable' and can be very useful, especially where the chemical hazard is low and the danger is from, for example, droplets or small splashes. The development of air-permeable materials has been so rapid that breathable CPC may become available for many duties in the not-too-distant future.

Air-impermeable materials, e.g., plastic- or rubber-coated, woven textiles and spun-bonded fabrics, plastic films, etc., do not permit penetration because they have no pores. However, permeation is a real possibility. Although items are sold as 'chemically resistant', many polymer coatings are permeable to some extent. Permeation resistance depends greatly on the type of polymer film or coating and the chemical or mixture of chemicals. Some CPC materials have very good permeation resistance to a wide range of chemicals but may be susceptible to one particular group of chemicals. Recent developments in multi-layer coated materials combine several different polymer films to give extremely high permeation resistance.

While laboratory data on permeation resistance should be obtained as an indicator of the performance of the CPC material, it should be remembered that such data will have been gathered from laboratory tests under specific testing conditions. For example, it may be possible to use a protective suit with low permeation resistance (breakthrough time) if the suit can be removed as soon as it is splashed. However, if the suit is splashed continuously or cannot be removed if it becomes splashed, a suit with a higher permeation breakthrough time should be used.

Some types of suits, e.g. air-supplied suits, allow small quantities of hazardous chemicals to enter the suit through penetration or permeation and are designed to 'sweep' these contaminants out of the suit by a flow of compressed air.

6.3 Styles of chemical protective clothing

Numerous styles and designs of CPC have been developed to meet the requirements of chemicals-based industries. To develop European standards for CPC, a classification system based on performance has been devised. This comprises the following categories:

— Gas-tight, protective clothing (Type 1),
— Air-supplied, non gas-tight protective clothing (Type 2),
— Liquid-tight, protective clothing (Type 3),
— Spray-tight, protective clothing (Type 4) and
— Partial cover garments (aprons, sleeves etc.)

Gas-tight protective clothing is sub-divided to account for different ways of including respiratory protection to this type of equipment, e.g. self-contained breathing apparatus worn inside or outside the suit and air-fed, gas-tight suits. Further European standards are being developed for limited-use, CPC and protective clothing for use against particulates. These will follow a similar structure.

Requirements given in the new European standards are designed to give as much information as possible to people selecting CPC. Tests required by these standards give a broad indication of garment performance across a range of key tests. The data is presented by the manufacturer in a range of performance levels and thus describes a performance profile for the garment (see Fig. 5).

6.4 Selection of chemical protective clothing

The selection is made by trying to balance the requirements identified by the workplace assessment, the performance data obtained from laboratory tests and other factors

	Low performance			High performance		
Abrasion Resistance	1	2	3	[4]	5	6
Coating Adhesion Strength	1	2	3	[4]	5	
Stability to Heat	1					[2]
Flex Cracking	1	[2]		3	4	
Puncture Resistance	1	2	3	[4]	5	
Tear Resistance	1	2	3	4	5	
Seam Strength	1	2	[3]	[4]	5	
Chemical Permeation						
Chemical A	1	2	3	[4]	5	6
Chemical B	1	2	3	4	[5]	6
Chemical C	1	2	3	4	[5]	6
Chemical D	1	[2]	3	4	5	6
Spray Test	FAIL					[PASS]
Jet Test	[FAIL]					PASS

Fig. 5. Typical performance profile for chemical protective clothing. Note: these performance levels are taken from laboratory tests given in European standards. Results within the boxes illustrate the test data for a typical item of CPC.

such as wearer comfort, ease of maintenance and costs, etc. With the development of standardized test methods and presentation of test data in a consistent manner, CPC can be selected in a logical fashion. Nevertheless, apart from the earliest stages of the selection procedure, the balancing of these various factors can be difficult and relies heavily on personal judgement.

For example, if the assessment identified the need to protect against a highly toxic and corrosive gas, in an emergency situation, it may seem appropriate to select a gas-tight suit offering high permeation resistance to the gas, almost certainly with self-contained breathing equipment worn inside the suit. However, consideration of other factors, such as abrasion resistance, tear resistance, suit weight, wearer comfort, etc., will also be important to ensure that the correct balance of performance is maintained. It may also be possible to use a limited-use suit with a defined life-span (say one hour maximum) rather than a re-usable suit which would have to be decontaminated each time before re-use.

It can be seen that selection is a careful balance of many factors, some of which may be in conflict, e.g. fabric weight and wearer comfort. Two or three options which satisfy the protection requirements should be identified and then offered to the wearers for them to select the most comfortable. Involving wearers in the final stage of the selection is more likely to result in acceptance by them.

6.5 Cleaning and maintenance of chemical protective clothing

If CPC has to be cleaned, two separate factors must be considered. Cleaning for hygiene reasons and decontamination must be treated separately. Whilst surface dirt and sweat, etc., must be removed after each use, decontamination may only be needed if chemical splashing has occurred. Decontamination can be defined as the reduction or neutralization of potentially harmful chemicals—from the surface and from within the materials of construction of CPC—to an acceptable level. Any decontamination procedure should be specific to the chemicals handled and must not damage the CPC material.

Most items of CPC cannot be successfully repaired. If a repair is to be successful, it must match or exceed the protective performance of the garment as a whole. For complex CPC, e.g. gas-tight suits, it may be possible to replace some component parts without difficulty. However, repairs to protective clothing materials are not usually effective.

An important function of anyone involved in the cleaning of CPC is visually to inspect the clothing for damage or deterioration. Records of inspections and, for more complex items, such as gas-tight suits, simple tests should be kept for each item of CPC (except for minor items such as gloves). The working life of CPC may have been specified during the selection process. Contaminated, time-expired or defective items should be condemned, rendered completely unserviceable and disposed of through a proper disposal procedure, taking into account the potential hazard from the chemically contaminated clothing.

7. TRAINING AND WEARER APPRECIATION OF PPE

Although assessments and selection of PPE may be carried out successfully, the use of PPE remains one of the most difficult areas to control and the most likely to lead to wearer exposure. Training should be focused on the needs of individuals and care taken to ensure that wearers know:

(a) **Why the PPE is needed**, what the hazards are and how the chemicals will affect them.
(b) **When the PPE is needed**. Whether the equipment is to be worn all the time or only in an emergency. How long the PPE can be worn. Whether it should be removed as soon as it is splashed or can be worn if contaminated.
(c) **How the PPE works**. The wearer should know how the PPE works, what its limitations are and how to recognise any faults which may reduce the protection afforded by the PPE.

(d) **Where the PPE should be worn**. Wearers must be aware that contaminated clothing should be removed near to hazardous areas to avoid spreading contamination to 'clean' areas. Wearers should also know the procedures for safe decontamination and disposal of used CPC and where to obtain replacements.

The intensity and complexity of the training should address the needs of wearers, practical training often being the best. Some form of simple test should be included to check how much knowledge has been absorbed. This can be achieved in practice periods or group discussions. After initial training, brief refresher courses should be given at regular intervals. Training records, indicating the course content and the identity of the trained personnel, should be kept. Some companies even 'license' their PPE users to show competence and restrict the issue of PPE to license holders.

8. PURCHASE AND SUPPLY

In many organizations, the purchase and supply of equipment is managed separately from those who use from those who use or maintain it. If this is the case, comprehensive descriptions or 'purchase specifications' need to be established to ensure that the correct equipment is supplied. Similarly, adequate stocks of equipment and spares should be kept to ensure that PPE is in sufficient supply and readily available.

9. CONCLUSIONS

The principle that PPE should always be regarded as a 'last resort', when other control measures (such as engineering controls or systems of work) cannot adequately control exposure to hazardous substances, is well established (although not always followed). Because PPE forms the 'last line of defence', it is vital that it performs properly in the workplace. Inadequate PPE or the inadequate use of PPE, may result in exposure to hazardous substances, possibly with fatal consequences.

The development of a successful PPE programme entails not only the selection of suitable PPE but also the provision of suitable training, maintenance, etc., to ensure that the equipment is used correctly and maintained in an efficient condition.

The first stage in the selection of PPE requires an assessment of the task and a clear understanding of the hazards and risks involved. The assessment should define the *real* consequences of the spill or exposure and should not rely solely on the material safety data sheet. The possibility of multiple hazards—chemical and non-chemical—should be considered as this will affect the PPE selection procedure. Consequently, all but the simplest assessments usually require a team approach and need to be reviewed frequently.

To select suitable PPE to satisfy the requirements of assessments, various types of PPE must be considered. A thorough knowledge of the advantages and disadvantages of each type is necessary to achieve a satisfactory level of protection in the workplace. The wide variety of PPE available may mean that several types of PPE may offer similar protection. The most important part of any PPE programme is the routine use of PPE in the workplace. The needs of the wearer must be considered thoroughly in the selection procedure as uncomfortable or inconvenient PPE will be worn incorrectly or not all. Wearer training and hazard appreciation are vital to the successful use of PPE; failure to wear PPE correctly is possibly the most likely cause of wearer exposure.

Development of a successful PPE programme depends on the careful consideration of various key factors including hazard and risk assessment, PPE selection, purchase and supply, training and wearer appreciation and maintenance and disposal. Failure to consider any one of these factors can result in an ineffective programme with the likely consequence that PPE users will be exposed to substances which present hazards to health.

REFERENCES

[1] 'Control of Substances Hazardous to Health Regulations 1988', SI 1988 No. 1657 (ISBN 0 11 0876571), 1988, HMSO.
[2] Health and Safety at Work etc. Act 1974 (ISBN 0 10 543774 3) 1976, HMSO.
[3] 'Occupational Exposure Limits' HSE Guidance Booklet EH 40, HMSO (Revised annually).
[4] 'Selection, Use and Maintenance of Chemical Protective Clothing', BS 7184, British Standard Recommendations, 1984, BSI.
[5] 'Minimum Health and Safety Requirements for the Use by Workers of Personal Protective Equipment at the Workplace', Directive 89/656/EEC, *Official Journal of the European Communities*, L 393, pp 18–28, 1989, HMSO.
[6] 'Approximation of the Laws of Member States relating to Personal Protective Equipment', Directive 89/686/EEC, *Official Journal of the European Communities*, L 399, pp 18–38, 1989, HMSO.
[7] 'Specification for Particle Filters used in Respiratory Protective Equipment', British Standard BS EN 143:1991, BSI.
[8] 'Specification for Gas and Combined Filters used in Respiratory Protective Equipment', British Standard BS EN 141, 1991, BSI.
[9] 'Recommendations for the Selection, Use and Maintenance of Respiratory Protective Equipment', British Standard BS 4275, 1974, BSI.
[10] 'Respiratory Protective Equipment—A Practical Guide for Users', HSE Guidance Booklet HS(G)53, (ISBN 0 11 885522 0), 1990, HMSO.
[11] 'Respirator Performance Terminology', H. P. Guy, *Am. Ind. Hyg. Assoc.*, May 1985, p B-22.
[12] 'Managing the Selection and Use of Chemical Protective Clothing', G. C. Coletta and M. W. Spence, *Performance of Protective Clothing*, ASTM STP 900, R. L. Barker and G. C. Coletta, Eds., American Society for Testing and Materials, Philadelphia, 1986, pp 235–242.
[13] 'Protection of the Eyes', Third Edition, 1990, CIA.
[14] 'Selection, Use and Maintenance of Eye-Protection for Industrial and other Uses', British Standard BS 7028, 1988, BSI.

10

Occupational exposure limits

Mr A. M. Moses

SUMMARY

The purpose of occupational exposure limits (OELs) and the various situations which lead to a need for them are described. The history of OELs is covered briefly, leading to a review of the different types of limit currently extant. Regulatory requirements and developments are described and a number of working definitions offered (health-based limit, working limit, airborne and biological limits, etc.).

The process which should be followed in establishing a health-based, airborne OEL is outlined. This is illustrated by a flowchart and covers the nature and extent of the initial database required, identification of adverse effects, assessment of the gravity of the effects, assessment of dose–response relationships, establishment of 'lead' effect(s) and identification of 'no adverse effect levels', leading to a recommended figure for a health-based OEL. Problems arising in this process are discussed and, in particular, the factors which should be taken into account in moving from a 'no adverse effect level' to a health-based OEL are covered.

The importance of validated sampling and analytical methodology is emphasized, as is the need to give careful thought to an appropriate reference period. Specific problems (e.g., ready skin penetration, respiratory sensitization) can be addressed by means of annotations.

The overall process of health-based OEL setting is one which requires certain competences and these are outlined. The overriding need for adequate and transparent documentation of the process is stressed.

A short section is devoted to ways in which OELs should be used and may be misused. Some of the difficulties arising from multiple exposure (mixtures) and non-standard working patterns are discussed.

The resource-intensive and time-consuming nature of the process described is considered against the background of the increasing need for OELs (or their equivalent) in order to demonstrate adequate control of exposure to new and existing chemicals in a variety of situations. Possible ways forward are briefly considered.

1. INTRODUCTION

By the early decades of this century, it had become widely accepted that excessive exposure to certain substances could, and did, cause adverse health effects in people who had to work with them; lead and mercury provide obvious examples. The legislative framework of the time, while recognizing this reality and indicating that governments had an interest in the health of working populations, was largely focused on the provision of compensation for victims of such adverse effects. It was not until the 1930s that the principle of preventing adverse health effects, by controlling exposure to substances in question, began to establish itself with the advent of the profession of industrial hygiene. In its early years, this discipline, which combined elements of engineering, chemistry and medicine, had some difficulty in gaining credibility. It had to contend with entrenched attitudes of management (who tended to regard such activities as counter-productive), labour unions (whose principal interests were in 'danger money' and compensation) and the defensive attitude of some of the long-standing professions involved.

In the early 1940s, two major factors came together to provide the genesis of the concept of occupational exposure limits. In the first place, the drive of the relatively new discipline of industrial hygiene, in seeking to apply the principle that 'prevention is better than cure', was searching for quantitative methods of demonstrating that exposures to chemicals were, indeed, adequately controlled. Secondly, analytical methodology and instrumentation had developed to the stage where it was practicable, albeit crudely, to measure exposure at the workplace. These naturally led to the need for quantitative criteria against which to judge whether a given measured exposure level was satisfactory and the concept of Threshold Limit Values (TLVs) began to be developed through the auspices of the American Conference of Governmental Industrial Hygienists (ACGIH) in the USA.

The following decades saw significant changes in attitudes, with management coming to realise that controlling occupational exposure was, in fact, less expensive than paying compensation and unions increasingly pressing for safer workplaces and more information. Legislation in the field has also developed, with today's laws requiring industry to provide demonstrably safe workplaces as well as to provide information on hazards, risks and methods of controlling risks from working with hazardous chemicals.

Against this background, a number of independent approaches to setting occupational exposure limits have developed in industrial nations. In the USA, the ACGIH TLV listing [1] has developed to cover some 800 substances, with Biological Exposure Indices (BEIs) for approximately 25 substances. The ACGIH TLV lists were, for many years, reprinted verbatim for use in the UK. A move towards an independent UK system began with the introduction of Control Limits and Recommended Limits following the Health and Safety at Work Act etc. 1974. The Control of Substances Hazardous to Health Regulations (COSHH) 1988 [2] led to the current system of Maximum Exposure Limits (MELs) and Occupational Exposure Standards (OESs). These MELs and OESs are established by the Advisory Committee on Toxic Substances (ACTS) and its dependent Working Group for the Assessment of Toxic Chemicals (WATCH). These bodies report to the Health and Safety Commission and take account of inputs from the Government, industry, trade unions, local authorities and independents. The limits are published annually by the HSE [3].

Within Germany, an independent system for establishing Maximale Arbeitsplatzkonzentrationen (MAKs) and Technische Richt konzentrationen (TRKs) has been in place for many years [4]. A separate system has also been developed within The Netherlands. Within the European Community, a Commission (DG V) initiative is under way which will lead to the promulgation of both Binding Limit Values (BLVs) and Indicative Limit Values (ILVs). It remains to be seen how these values, particularly the latter, will be integrated into existing national systems.

The former Soviet Union has developed a system of Maximum Acceptable Concentrations (MACs) which are, in many instances, set at levels much lower than their counterparts in western countries. Their credibility, as practically applicable limits, has however, been called into question. Many other countries publish lists of occupational exposure limits, often based on the ACGIH TLVs of the USA.

2. CURRENT SITUATION IN THE UK

The COSHH Regulations came fully into force on 1st January 1990. The Health and Safety Commission, which was responsible for preparing the Regulations, and which, through its executive arm, is responsible for enforcing compliance, has described COSHH as the most far-reaching regulations it has introduced since coming into being in 1974.

The COSHH Regulations apply to all situations involving employees and substances in the workplace. The purpose of the regulations is to protect workers' health by requiring substances in the workplace to be identified and the potential risks to health associated with the use of those substances to be assessed. Either directly, or through associated codes of practice, the regulations lay down a wide range of duties for both employers and employees for action to control risks to employee health. The regulations are, therefore, of great significance to the chemical industry in which a principal activity is the synthesis and processing of substances (see also Chapter 1).

COSHH requires exposure to hazardous substances to be prevented or, if this is not reasonably practicable, adequately controlled.

For substances appearing in Schedule 1 (for which Maximum Exposure Limits (MELs) have been set), exposure must be reduced as far as reasonably practicable and, in any case, below the MEL. For substances with an approved Occupational Exposure Standard (OES), the level of exposure should not exceed the standard unless the employer has identified the reasons for exceeding the standard and is taking appropriate steps to comply with the OES as soon as reasonably practicable. Lists of MELs and OESs are reproduced in the current edition of the Health and Safety Executive (HSE) Guidance Booklet EH 40 (revised annually) [3].

For substances which have not been assigned official exposure limits under COSHH, the COSHH general Approved Code of Practice, paragraph 29, gives practical advice that: 'In some cases there may be sufficient information to set a self-imposed working standard, e.g. from manufacturers and suppliers of the substance, from publications of industry associations, occupational medicine and hygiene journals'.

This reflects the practice that has developed over the years in several large companies of setting their own 'in-house' standards, where none exist in official lists and a need for such standards is perceived.

The setting of an occupational exposure limit is an expert task requiring the participation of occupational health professionals (e.g. occupational physicians, toxicologists and occupational hygienists). The following sections outline the process which should be followed and discuss some of the difficulties involved.

3. DEFINITIONS

It is apparent, from the preceding sections, that there is a proliferation of terms, abbreviations and acronyms in this field. Most of these stem from individual national systems and the following definitions are offered in an attempt to place the subject on a more rational basis.

3.1 Occupational exposure limit (OEL)
This is offered as a generic term to cover all situations in which a numerical value (usually a concentration of a substance in air or a biological fluid) is used as a criterion against which exposures can be judged in assessing whether control of exposure is adequate.

3.2 Health-based limit
While all OELs are intended to protect health, the phrase 'health-based limit' is offered to cover OELs which are derived solely from a directly health-related database and which take no account of whether the limit so derived can be achieved in practice.

An *airborne*, health-based limit may be defined as the maximum concentration of a substance, present in the breathing zone of workers, averaged over a specified reference period, to which, according to current knowledge, employees may be exposed day after day without adverse effects.

A *biological*, health-based limit may be defined as the maximum concentration of a substance (or a metabolite, as appropriate) which, when measured by a validated biological monitoring technique in an appropriate biological material (e.g. body fluids, expired air), will be associated with no adverse effects.

3.3 Working limit
A 'working limit' takes into consideration the practicability of compliance. Such a limit may be set where the nature or extent of the database precludes the establishment of a health-based limit. Although not solely health-based, any working limit must be

established at a level below that which carries a 'just tolerable' risk to health, as indicated by the database. In principle, different working limits may apply to the same substance in different working environments.

3.4 Action level

In some circumstances, an 'action level' may be set locally based on any of the above standards. Such a level is usually designed to trigger a specific risk management procedure appropriate to a particular situation and is not, in itself, an exposure limit.

An attempt is made in Table 1 to relate some of the better known types of official limits to the above definitions.

Table 1. Occupational exposure limits

Country	Health-based limits	Working limits
UK	Occupational exposure standards (OESs)	Maximum exposure limits (MELs)[a]
USA	Threshold limit values (TLVs)	
	Biological Exposure Indices (BEIs)	
Germany	Maximale Arbeitsplatz-konzentrationen (MAKs)	
		Technische Richt Konzentrationen (TRKs)
	Biologische Arbeitsstoff-toleranzwerte (BATs)	

[a] MELs in the UK cover some substances which fit the definition of 'working limit' given in this chapter, as well as others for which a health-based limit (OES) could have been established but with which limit it is not reasonably practicable for industry to comply. In these latter cases, the tripartite Advisory Committee on Toxic Substances (ACTS) have set a MEL at a level above that at which an OES could otherwise have been set.

4. PROCEDURE FOR SETTING IN-HOUSE OCCUPATIONAL EXPOSURE LIMITS

The overall procedure recommended for the setting of in-house OELs is summarized in the flowchart in Fig. 1 [5].

In the first instance, a clear need for an OEL should be established before embarking on what is always a time- and resource-consuming exercise. Such a need may arise for any substance which may contaminate the atmosphere when produced or used. For example, an occupational hygiene workplace assessment may indicate that an OEL is needed as part of the process of ensuring that there is adequate control of

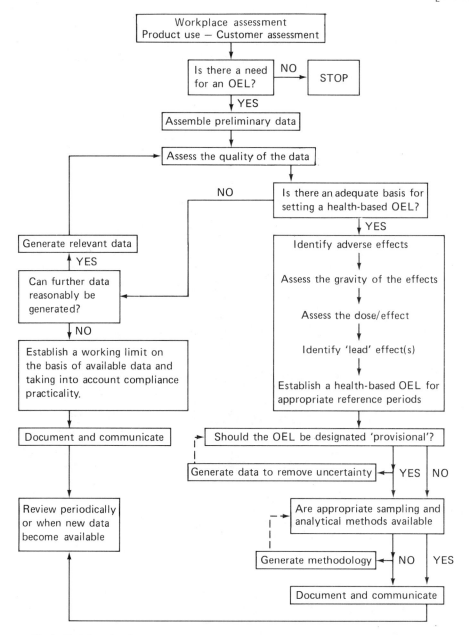

Fig. 1. Flowchart for the procedure for setting in-house occupational exposure limits (OEL).

risk. Alternatively (or additionally), OELs may be required for products where the nature of their use by customers is such that exposure by inhalation will occur and a valid standard is needed to ensure adequate control of risk on his premises. Although there may be exceptions, there is generally little point in establishing OELs unless the exposure situations call for a programme of exposure monitoring to establish adequate control of risk. Conversely, there is little point in extensive monitoring if there is no OEL to use as a standard.

4.1 Collection of data

Having decided that an attempt should be made to set an OEL, the next step is to gather together all the relevant data. It must be remembered that, at the end of the exercise, the process leading to the OEL should be adequately documented such that the logic is transparent. Information should be collected in the following areas:

— name of substance, synonyms, chemical structure, composition, principal contaminants;
— chemical and physical properties (which must include solubility data and, where relevant, volatility and/or particle size distribution data), stability and reactions in air;
— the names of chemically, physically or biologically analogous substances for which health-based OELs already exist, plus supporting documentation;
— toxicological data on the substance plus any epidemiological data and information concerning accidental over-exposure or poisoning in humans;
— metabolic and pharmacokinetic data—essential when setting a biological OEL.

A literature search should be carried out to ensure that all relevant information has been retrieved. References should be listed, numbered and annotated in an appropriate fashion. Against each, it should be noted whether the information comes from an abstract, summary, full report or publication. Wherever possible, original publications should be obtained and reviewed.

4.2 Evaluation of the data

At this stage, the quality of the overall database should be assessed to establish whether there is a sufficient basis from which to establish a health-based OEL. This is a complex question and ultimately one for expert judgement. Nevertheless, some broad guidelines can be given. Human data naturally take precedence over equivalent animal data and may, in some circumstances, provide a sufficient basis. Anecdotal information indicating that no problem has become apparent over a given period of time should be treated with caution. The temporal relationship between exposure measurements and any adverse health effects reported in a study should be carefully considered. Many field studies, particularly older ones, suffer from inadequate characterization and quantification of exposure.

In the absence of human data, animal data may well provide an adequate basis. In general, such data should include the results of at least one well-conducted, repeated exposure (sub-acute or longer) study by the inhalation route. Acute (single exposure) data alone are, in general, inadequate. Repeated exposure studies by the oral route are more generally available than inhalation studies and may well provide valuable data. However, it must be appreciated that any local effects on the respiratory tract will not be covered. Such studies should be accompanied at least by an acute inhalation study. Many factors will be taken into account by the toxicologist in assessing the adequacy of a repeated exposure study. These will include age of study, species, route of administration, exposure regime, controls, test substance characterization, method of atmosphere generation and analysis, extent of clinical and other observations, extent of gross and histopathology, choice of exposure levels and extent and quality of reporting. An ideal sub-acute inhalation study might cover three exposure levels, with clearly definable adverse effects at the top dose, less severe but still identifiable manifestations of the same effect at the middle dose and no effects seen across a wide range of parameters at the lowest dose. The subject of the collection and evaluation of toxic hazard data is covered more fully in Chapter 6.

Setting OELs by analogy with other substances is an exercise fraught with difficulty and pitfalls and should only be conducted with the assistance of an experienced toxicologist.

If, on this assessment, the database is considered inadequate but could be made adequate by the generation of relevant data, a preferred option may be to generate the missing data, if this can reasonably be done. In the situation that it is not reasonable to generate an adequate database, a decision may be taken to establish a working limit.

4.3 Qualitative hazard assessment

Given an adequate database, the next stage is to move on to a qualitative assessment of the hazard. The aim of this part of the assessment is to establish a list of biological effects resulting from exposure of animals or man which need to be taken into account when establishing an OEL. All adverse effects noted in humans should be considered. Each adverse effect seen in animals should be considered individually and a decision taken as to whether it is likely to be expressed in humans. The gravity of each effect should then be judged. The following scale may be used:

- **severe** —this category includes all causes of premature death, deformity and irreversible changes;
- **moderate**—changes which are easily detectable and which resolve on cessation of exposure;
- **minor** —changes of little or no clinical significance which do not progress and which resolve promptly on cessation of exposure.

The quality of the data should be assessed by appropriate experts (e.g. toxicologists, epidemiologists, occupational physicians, hygienists and analysts). This is particularly important in the case of data which may play a leading role in the establishment of an OEL. Clear reasons should be recorded for disregarding any particular item of data as should the basis for reaching decisions from contradictory evidence.

4.4 Quantitative assessment of risks

Having assessed critically the quality of the database and assembled a list of adverse effects for further consideration, the next stage is to conduct a quantitative assessment of the likely risks to humans presented by each of these effects. This starts with an

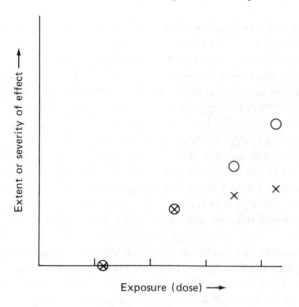

Fig. 2. Dose–effect relationships. ○, 'steep' dose–effect; ×, 'shallow' dose–effect.

examination of the data relating to each effect to establish whether a clear dose–effect relationship has been established. The relationship between dose (or exposure) and extent or severity of effect is critical in determining the highest exposure level which is likely to be without adverse effect (the 'no-effect level'). The principle is illustrated in a stylized way in Fig. 2. In this figure, two sets of data indicating both a 'steep' and a 'shallow' dose–effect relationship are shown. In this hypothetical example, both substances show the same 'no-effect level' but the consequences of over-exposure may be more severe for the substance with the steep dose–effect curve. This observation will be taken into account in setting an OEL. If, however, there had been no demonstrable 'no-effect level' for either substance, it might have been inferred from the slopes of the curves that the substance with the shallow dose–effect would show a lower 'no-effect level' than the substance with a steep dose–effect, again with implications for the levels at which OELs would be set. This illustrates the complexities of the process, since datasets in practice will rarely be ideal, and stresses the need for expert toxicological judgement. If dose–effect relationships are not clearly established but can reasonably be inferred, e.g. by analogy with other substances, the reasons for any such inference should be clearly stated.

It may well be that a particular substance has been investigated in more than one animal species, with a wide spectrum of possible outcomes (different effects, different severities, different dose–effect relationships, different 'no-effect levels'). In interpreting the results of such a set of studies, the basic guiding principle should be that, in the absence of evidence to the contrary, humans are assumed to be at least as sensitive as the most sensitive species tested. If it is decided to discount observations seen in any particular species as irrelevant for humans, the case for doing this must be soundly based scientifically and adequately documented. Such a case may arise, for example, if there is adequate evidence of different metabolic pathways for the substance in humans and, say, mice and the adverse effect in question can be clearly attributed to a mouse-specific metabolite. Concordant results in more than one animal species will increase the level of confidence in establishing a 'no-effect level'.

4.5 Determination of a no-effect level
Once the list of adverse effects has been reviewed along the lines described above, it will become clear which effect (or effects) will be of most significance in determining the level of an OEL (the 'lead' effect). It is particularly important to obtain the key articles, reports or publications relating to this 'lead' effect and to subject them to a detailed expert review, the outcome of which should be documented.

This process should lead to the establishment of a 'no-effect level' in the species in question. This may be a human 'no-effect level' (if the key data relate to human exposure and effects) or it may be an animal 'no-effect level'. Particular difficulties may arise where the 'lead' effect is respiratory sensitization or carcinogenesis with a genotoxic mechanism of action. The mechanisms involved in both these effects suggest that, from a theoretical standpoint, it may not be possible to establish an exposure level below which there is unlikely to be any effect. Thus establishment of a true 'health-based' OEL may not be practicable. In such cases, a 'working limit' may be established. Substances exhibiting these effects may be accorded MEL status (in the UK) or TRK status (in Germany). For carcinogens with a non-genotoxic mechanism of action, it may well be possible to establish a threshold of effect and a 'no-effect level', thus leading to a true 'health-based' OEL.

Another situation which may lead to the establishment of a 'working limit' is that in which the database is inadequate to establish a 'health-based' limit. In such cases, the work needed to progress to a health-based limit should be defined and the practicability of conducting such work considered.

4.6 Relationship between no-effect level and the OEL
Assuming that an adequate database exists and that it has proved possible to establish a 'no-effect level' with a reasonable degree of confidence, the final stage in the process of setting a 'health-based' OEL is to determine the relationship between the 'no-effect level' and the OEL. This is not a straightforward exercise and is not amenable to 'rules of thumb'. It requires multi-disciplinary expert consideration which should take into account at least the following factors:

— the scientific confidence in key studies,
— whether the database is human or animal,
— the severity of the 'lead effect',
— whether the 'lead effect' is well understood (or rare) in humans/animals,
— the extent of concordance in animal studies,
— whether the 'no-effect level' is established directly or inferred,
— the slope of the dose/effect curve,

- known species differences,
- local or systemic effect,
- whether the effect is due to the parent molecule or a metabolite,
- pharmacokinetic data (e.g. half-life) and
- precedents established from other substances.

As part of the process of setting a 'health-based' standard, an appropriate reference period should be established. This will often be 8 hours (i.e. a working day or shift) but, where the 'lead effect' is acute (e.g. respiratory irritation), it may be more appropriate to set a 10-minute reference period.

The entire process outlined in this section should be documented to an extent sufficient to make the rationale underlying the establishment of an OEL clearly evident to other professionals. Once the OEL has been established and adequately documented, the outcome and rationale should be adequately communicated to all those with an interest in the issue.

5. OTHER FACTORS ATTENDING THE ESTABLISHMENT OF AN OEL

An OEL should be accompanied by validated atmospheric sampling and analytical methods. As stated earlier, there is little point (in a working situation) in putting effort into establishing an OEL and then not being able to use it in practice. In the absence of such methods, steps should be taken to develop them in advance of needing to use the OEL in practice. An OEL without such attendant methodology may be useful, however, at the design stages of a new plant or process, but will be of little value once actual operations commence.

During consideration of the toxicological properties of substances, the possibility of biological or biological effect monitoring should be considered. This may, in appropriate circumstances, lead to the establishment of a biological 'health-based' limit as well as, or instead of, an airborne limit. Such limits are less common than airborne limits and the process for their establishment is not described in detail in this chapter although the principles outlined in section 4 above will apply. Biological limits have the advantage of being able to be used to assess the adequacy of exposure control by all routes (inhalation, skin and oral) and, given an appropriate database, may be of particular value where a substance is absorbed through the skin to a significant extent or where non-occupational exposure to the same substance may be important (see also Chapter 12).

Also, during consideration of the toxicology, the likelihood that unduly susceptible individuals or groups may occur in the exposed population should be considered. In the latter case, it may be necessary to develop, as an adjunct to the OEL, a system for identifying (screening) hypersensitive individuals. Where the occurrence of hypersensitive individuals is likely to be other than rare, the OEL should be established at a level sufficiently low to protect them.

Where an airborne OEL is being established and there is evidence that dermal penetration of the substance may contribute significantly to overall exposure, it must be made clear (e.g. by accompanying the OEL with the word 'skin') that the OEL is only valid where adequate precautions have been taken to prevent dermal absorption. All OELs should be reviewed periodically or when new data become available.

6. APPLICATION OF OELs

It is not the purpose of this chapter to review in any depth the application of OELs in practice as this is a major topic in its own right. Subjects such as appropriate sampling and analytical methodology and how these should be used to determine compliance (or conformance) with an OEL are the province of occupational hygienists (see Chapter 11). Informed use of an OEL, in practice, requires an understanding of its toxicological basis; two examples of this are briefly discussed below.

6.1 Mixtures

In practical situations, exposure to more than one substance is the rule rather than the exception and most OELs are set with exposure to a single substance in mind. When using OELs in mixed exposure situations, it is important to understand the nature of the biological interaction of the substances involved in the body. Such interactions fall broadly into three categories which will determine the way in which OELs should be used:

— *independent substances*: where the 'lead effects' are clearly biologically independent, (e.g. respiratory irritation and liver damage), it will be sufficient to ensure compliance with each OEL individually;
— *additive substances*: where the 'lead effects' are the same or similar, (e.g. both substances cause broadly similar liver damage), the effects of exposure to both should be assumed to be additive and compliance with the OELs assessed accordingly. Such an interaction should be assumed unless there is evidence to the contrary;
— *synergistic substances*: this category covers mixtures of substances where the total effect on the body is greater than the sum of the individual effects. Such an interaction is rare but specialist toxicological advice should be taken if such interactions are suspected.

The above considerations apply to contemporaneous exposure; where exposures are consecutive (e.g. moving from one plant to another in the course of a working day), it may be appropriate also to take into account the pharmacokinetics of a substance. If, for example, a substance is cleared from the body within a few minutes (and the parent substance is responsible for the 'lead effect'), it may be appropriate to regard as independent, consecutive exposures to two substances for which contemporaneous exposure would be additive (See also Chapter 6, p. 102.)

6.2 Non-standard working patterns

OELs are conventionally established to relate to an 8-hour working day (or shift). In practice, many work schedules involve working for longer periods (e.g. 12-hour shifts) and the question arises of how to relate an 8-hour OEL to such a work period. The simplest (and most cautious) approach is to reduce the OEL pro-rata to the length of shift compared with 8 hours and this should be adopted in the absence of other information. However, a knowledge of the pharmacokinetics of the substance in question may permit a different approach. For substances with either a very short or very long biological half-life (relative to 8 hours), it may be more appropriate to use effective reference periods much shorter (e.g. 10 minutes) or much longer (e.g. a 40-hour working week) when determining compliance. Such considerations are complex, fraught with difficulties and should not be attempted without expert involvement.

7. SOME COMMON MISUNDERSTANDINGS

7.1 — 'It doesn't have an OEL, so it must be safe'

Whether a substance has an official OEL is no indication of the extent of the hazards of the substances or of the potential risks which exposure to that substance may entail. Official OELs exist for relatively few substances and reflect the practical need for control standards rather than anything else. It may be more appropriate to take the opposite view, namely that a substance with an OEL is easier to handle and control safely than one which has not been so assessed and for which the hazard and risk may be unknown.

7.2 —'It has a nasty smell, so there is no need to set an OEL'

Two factors should be borne in mind here. Firstly, smell thresholds usually bear no relationship to toxic thresholds. Although some substances may have smell thresholds well below toxic levels, the reverse can easily be the case. Secondly, even for the former group, relying on smell as a warning property is unsatisfactory due to wide inter-personal variations in smell thresholds and the phenomenon of 'smell fatigue'.

7.3 —'I'll change from X with an OEL of 20ppm to Y with an OEL of 50ppm; Y must be safer'

While it is sound commonsense and good occupational hygiene practice to replace toxic substances with less toxic ones wherever practicable, such decisions must be taken on an informed basis and not on the basis of numerical comparisons alone. Other factors which should be taken into account include the nature of the lead effects, the physical properties (volatility, dustiness, etc.) and the flammability and explosivity of the substance.

7.4 —'Substance X with a MEL is more dangerous than substance Y with an OES'

While some substances allocated MELs may well be genotoxic carcinogens, respiratory sensitizers or other highly toxic substances, which clearly need treating with appropriate respect, this is not always the case and the documentation underlying the decision to set a MEL should be examined. In some cases, this may indicate that a 'health-based' OES could have been set but a numerically higher value (a MEL) was necessary if certain uses were to be reasonably practicable. Simplistic comparisons of this nature should not be made.

7.5 —'I want to use substance X; it hasn't got an OEL so would you please calculate/deduce/look one up for me in the next week'

It is quite clear, from earlier sections of this chapter, that establishing an OEL is a complex, time- and resource-consuming exercise which cannot be carried out in a hurry and should not be embarked upon superficially or lightly.

8. FUTURE DEVELOPMENTS

There is increasing pressure (through COSHH in the UK and similar workplace legislation elsewhere) to accelerate and extend the process of setting OELs to ensure that control regimes are adequate to protect health. The scale of the problem is made apparent by the realization that EINECS (the European Inventory of Existing Commercial Chemical Substances) covers more than 100 000 substances, for which fewer than 1000 have established OELs. It is apparent that the database required for, and the process involved in, establishing a health-based OEL are both such as to preclude the establishment of OELs for all substances of concern within a reasonable time frame.

Ways of simplifying the process are continually being sought, both by official standards-setting bodies and private industry. These will undoubtedly lead to a speedier process with a higher quality product but it may be that a complementary initiative is required.

In many practical situations, a broad indication of an acceptable range of exposure is all that is needed to drive decisions between one exposure control regime and another, and that over-sophistication may, in many cases, be unnecessary.

There may thus be a case for developing a simple method of assigning substances, on the basis of hazard, to one of a limited number of categories which can then be generally used, in the absence of other information, as a basis for assessments leading to selection of exposure control regimes. Such a method would, in no way, replace the process for establishing health-based OELs, where a need for such limits is apparent and an adequate database exists.

REFERENCES

[1] '1991–1992 Threshold Limit Values for Chemical Substances and Physical Agents and Biological Exposure Indices', 1991, ACGIH. (Revised annually).
[2] 'The Control of Substances Hazardous to Health Regulations 1988', SI 1988 No. 1657, 1988, HMSO.
[3] 'Occupational Exposure Limits 1992', HSE Guidance Booklet EH 40/92 (ISBN 0 11 885696 0), 1992, HMSO. (Revised annually).

[4] 'Maximum Concentrations at the Workplace and Biological Tolerance Values for Working Materials 1991', No. 28, 1992, Verlag Chemie. (revised annually).
[5] 'Guidance on Setting In-House Occupational Exposure Limits (Regulation 7)', 1990, Chemical Industries Association.

11

Sampling strategies

Dr I. G. Guest

SUMMARY

This chapter provides practical guidance on the design and implementation of suitable sampling strategies for determining concentrations of hazardous substances in workplace atmospheres. Sampling may be used as part of the following processes:

(a) health risk assessment or epidemiological studies,
(b) evaluation of compliance/conformance with limit values,
(c) investigation of engineering and process control measures, and
(d) evaluation of trends in facility performance.

A structured approach to strategy design is set out which is based on the requirement to develop clear objectives for any measurement exercise. In this way, a maximum amount of useful information may be obtained in a cost-effective manner.

The choice of a strategy is affected by various other factors which are described in detail. These include temporal and spatial variation in the concentration of an airborne contaminant, the availability of suitable sampling and analytical methods, whether a substance exerts chronic or acute effects (or a combination thereof), mixed exposures and financial and manpower constraints. Overall, the resources devoted to monitoring should match the degree and nature of the risks to health which may arise.

Statistical models used for occupational exposure data are described together with their relevance to strategy design and decision making. Application of statistical concepts to sampling can result in a requirement for large numbers of samples. Techniques to mitigate this are described together with the consequences of failure to obtain a sufficient number of samples.

The interpretation of results is discussed together with the types of action that may be appropriate.

Exposure to substances hazardous to health may occur by inhalation, ingestion or skin contact/absorption. Monitoring of workplace atmospheres provides little or no information about the potential for exposure by the latter two routes. Integration of a programme to monitor airborne contaminant concentrations with other monitoring programmes, e.g. biological or biological effect monitoring or health surveillance, is briefly discussed.

1. PURPOSE AND GENERAL CONSIDERATIONS

Any occupational hygiene survey (whether of a qualitative or quantitative nature) is usually part of a decision-making process. It is, therefore, important to ask the fundamental question: 'Why are the data needed'. The answer to this question will define the aim or aims of the survey. This may seem obvious but experience indicates that all too often measurements are carried out without any clear purpose. The result is the collection of inappropriate data which is a futile and costly exercise which may also cause a false sense of security to develop.

The most appropriate strategy to employ, in a given situation, will depend upon the objectives of the exercise, which may include:

(a) health risk assessment (e.g. as part of a COSHH assessment),
(b) determination of compliance/conformance with occupational exposure limits (OELs),
(c) evaluation of control measures or plant performance,
(d) provision of data for epidemiological purposes,
(e) monitoring the continued performance of control measures, and
(f) validation or comparison of measurement methods (sampling and analytical methods).

For example, measurements to assess employees' exposures, in an epidemiological study, are done to establish a dose–response relationship for the hazardous substance being studied. Employees with a wide range of exposures will, therefore, need to be included in the survey. Testing for compliance with an OEL or making a health risk assessment, when an OEL is available, can be undertaken with far fewer measurements. Workplace inspection can eliminate many employees from the initial measurement programme if their exposures are obviously well below the OEL. Similarly, for employees with exposures obviously well above the OEL, the most appropriate action may be first to improve workplace controls and then measure to assess their effectiveness.

Any workplace sampling is essentially a synthesis of various elements which may be qualitative or quantitative. For example, a health risk assessment cannot take place without some type of qualitative assessment. The process requires some judgement about the potential routes of exposure from air, skin contact or ingestion and the various factors leading to potential exposure in the workplace. Measurement of airborne contaminant levels alone can never answer any of these questions satisfactorily. It, therefore, follows that the measurement of airborne contaminant levels and their comparison with OELs cannot be used as the sole measure of occupational health performance. Also, the results are not a yes/no indicator of whether a health risk exists. Neither the measurement results nor the standards to which they are generally compared are absolute so that care must always be exercised when interpreting data.

Measurement cannot be used as a substitute for the general process of decision making affecting the control of risks to health or for continually deferring such decision until more data is available. Measuring the concentrations of substances in the air itself does not control or reduce risks to health.

2. A STRUCTURED APPROACH

2.1 Factors to consider

In addition to the basic objectives of any occupational hygiene survey, which may require measurements to be made, a number of other factors can affect the strategies and protocols which will need to be developed. These can include:

(a) the nature of the airborne contaminant and associated hazards,
(b) the availability of suitable sampling and analytical methods,
(c) the type of process and the circumstances under which emissions may occur, and
(d) spatial and temporal variations in concentrations.

The last factor is of particular importance since it implies:

(a) a preference for personal sampling when trying to assess personal exposures, and
(b) a need for a statistical approach to the measurement strategy.

Instantaneous concentrations of air contaminants emitted from a plant or process in a workroom can vary from point to point within the room and will vary with time during the course of the workshift. In addition, the average concentration for either a single point in a workroom or the average for a number of points varies from shift to shift. Within- and between-shift variations can be due to combinations of factors such as:

(a) variation in the number of emission sources,
(b) variation in the rate of emission from a given source, e.g. due to changes in the process, and
(c) variation in the dispersion of a contaminant from its source due to changes in ventilation patterns.

The concentrations of a contaminant inhaled by employees are governed by the above factors. Additional factors which also contribute to the variations in the individual employee's exposure include:

(a) the employee changing position relative to the various emission sources,
(b) variations in tasks between employees, and

(c) variations in working methods between different employees undertaking the same task.

2.2 General planning

The setting up and carrying out of a measurement exercise to achieve a defined aim, therefore, needs to be broken down into a series of interdependent stages, each with its own objectives. There will be an overall, conceptual strategy within which individual elements will have their own strategic considerations. The advantages of such an approach include:

(a) the provision of a basis for planning and costing,
(b) development of a firm, expanding information base to assist in defining possible further priorities,
(c) flexibility to allow changes of direction, should this prove necessary, and
(d) the systematic acquisition of information which may provide early answers to the basic questions, saving the expense of conducting all stages of the exercise.

Structured approaches to sampling have been formalized into guidance or standards in a number of countries including the UK [1], USA [2], Germany [3] and France [4]. Increasingly, there is a move towards supra-national standards including a draft CEN standard [5] and earlier reports by CEFIC [6] and WHO [7].

The objective of these various standards and reports is usually to provide guidance on the development of measurement strategies in relation to compliance with national legislation or OELs. The various proposals have similar principles based on a three- or four-stage strategy as follows:

(a) initial appraisal,
(b) basic survey,
(c) detailed survey, and
(d) routine survey.

The first two stages are often combined in three-stage strategies.

As an example, in the UK, the Control of Substances Hazardous to Health Regulations 1988 (COSHH) require employers to make an assessment of the risks to health created by the work being undertaken and of measures required to protect their employees. Such an assessment under COSHH will certainly involve an initial appraisal and can involve a basic or detailed survey. There are also specific requirements under COSHH for routine monitoring, in some circumstances, and the fourth element is then relevant.

Some general factors affecting the choice of measurement strategy are outlined above. Occupational hygiene data often have a high level of variability so that any statistically-based strategy may require a large number of measurements to be made. This is discussed, in detail, below.

A well-planned strategy, with a large number of measurements, should provide great confidence in the results. This may not always be essential depending upon the circumstances and objectives of the survey. In practice, there are also limits to the resources available in terms of time, equipment, manpower and cost.

It is a primary duty of employers to ensure a healthy working environment for their employees and the employer needs sound and timely advice where improvements in control measures are required. Adequate data may be required to support a case for major capital spending but it is equally important to avoid excessive measurement which can delay the implementation of any improvements. However, if measurements are part of an epidemiological study, this may require a large number of measurements spread over a considerable period of time.

2.3 Initial appraisals

The appraisal of all available relevant information is an essential first step in any survey to determine whether measurements are required and, if so, where they should be targeted. In many cases, it may provide an adequate assessment of risk and allow decisions to be reached without incurring the cost of a quantitative survey. The decision may be that conditions are satisfactory and no action is required or that direct action to improve the standard of control is required. (Hazard evaluation and risk assessment procedures are covered in Chapters 6 and 7, respectively.)

The initial appraisal should answer the basic questions of 'what?', 'who?', 'when?', 'where?' and 'how?'. At the first stage, it is important to identify the potential routes of exposure (inhalation, ingestion, skin contact/absorption). Many surveys have been set up in the past to evaluate the potential for inhalation exposure where the real problem was related to skin contact and, less frequently, ingestion.

Where inhalation may be important, qualitative tools, such as the 'dust lamp' or smoke tubes, are useful as they allow a preliminary evaluation of the effectiveness of controls to be made.

2.4 Basic surveys

A basic survey may be adopted where limited quantitative information is required. Circumstances which require such an approach may include:

(a) start up of a new process,
(b) substantial changes to an existing process, operation or controls,
(c) unusual, infrequent or intermittent processes or operations, e.g. maintenance, plant decommissioning, cleaning or infrequent batch processing,
(d) an OEL has been set for the first time,
(e) a range-finding exercise as a preliminary to a detailed survey, and
(f) the initial appraisal indicates that an exposure problem exists.

The approach is also suitable where very poor control is suspected or extremely good control anticipated. Arguably, these situations should not warrant any quantitative measurements but, for various reasons, either employers or employees may require the reassurance that measurements are being done.

Before any measurements are made, any existing data should be considered as this can eliminate the need for additional measurements. For example:

(a) **Earlier measurements from the existing plant or process.** These need to be used with care as changes may have occurred which might affect exposure. Also, the measurement methods may not meet current standards.
(b) **Measurements from comparable plant, processes or workplaces.** Careful interpretation of the data is essential. Differences in plant may be obvious but more subtle differences may exist, e.g. work practices, such as job sharing, break patterns, etc. may alter the degree and pattern of exposure. For some processes, e.g. maintenance, conditions are frequently so variable that sites are rarely comparable.
(c) **Calculation.** Occasionally, this is a viable option, e.g. it has been applied to hydrocarbon distillate vapour composition [8].

A basic survey may employ relatively simple measurement methods, e.g. detector tubes, organic vapour analysers, aerosol/particulate monitors or gravimetric techniques. It is essentially a range-finding exercise in which it is usually most efficient to target the 'worst case' exposures, perhaps using these as an indicator of overall risk. Such an approach is heavily reliant on the skill of the person responsible for the survey and interpretation of the limited data. Inexperience may result in measurements for the wrong employees, at the wrong time or misinterpretation of the data. This can result in a health risk remaining unidentified or, alternatively, excessive costs being imposed on an employer to control a very limited problem.

Before any measurements are made, the pattern of exposure must be determined. If the pattern is extremely variable or large numbers of workers are involved (making it difficult to identify the 'worst case' situations), a basic survey may not be very reliable. Grouping of employees (see section 6 below) into homogeneous exposure groups may help, following a consideration of work patterns, job, duration of exposure, etc. Those groups with the highest predicted exposure should then be selected.

Measurements should be made over an appropriate period and this will depend on whether exposure is continuous or intermittent and whether peak exposures occur. It is essential that events at the start and finish of a work period are considered, such as weighing, reactor loading/discharge, cleaning, etc. Measurements will need to cover periods which make the major contribution to exposures. Ambient conditions should be taken into account, e.g. seasonal effects or wind speed and direction (outdoor work) together with possible differences between day and night shifts or the various stages of rolling shifts.

Personal sampling techniques are preferable for estimating personal exposures but, as noted earlier, simple detector tube or instrumental methods may be adequate. Static sampling may be useful in verifying the existence and magnitude of emissions or performance of control measures.

Sufficient measurements must be made to give confidence that, within reason, all relevant work patterns and exposure cycles have been covered. In addition to the 'worst case' employees, there is merit in including some employees with expected lower exposures. This provides some quality control over the survey and initial appraisal.

Bearing in mind that the basic survey is primarily a range-finding exercise and that limited data are available, decisions have to be made. This may include comparison of the results with an OEL. The data are unlikely to be suitable for compliance testing procedures (outlined in section 9 below) and there are no universal criteria for making decisions from a basic survey. Subject to the application of good sampling judgement, the possible decisions are:

(a) *Do no further sampling.* Low results indicate that a detailed survey is unlikely to be necessary. [Experience suggests that, if results are approximately 0.1.–0.2 × the OEL, the OEL is unlikely to be exceeded.] This is not a decision to do nothing. The survey may have revealed deficiencies in control which are open to improvement. Where a substance has a Maximum Exposure Limit (MEL), there is a duty to reduce exposures to as low a level as is reasonably practicable.
(b) *Carry out immediate investigative/remedial actions (unless there are good reasons to be suspicious of the results).* [A good proportion of the results are above the OEL.] Remonitoring may be required following the implementation of improved control measures.
(c) *Undertake a detailed survey.* [Results in the range 0.1–1.0 × the OEL, with isolated results above the OEL.]

2.5 Detailed surveys
The decision to conduct a detailed survey may derive from:

(a) the initial appraisal suggesting that the extent and pattern of exposure cannot be reliably assessed by a basic survey,
(b) the basic survey revealing that exposure is very variable, that large numbers of people are potentially at risk or that measured exposures suggest that OELs may be exceeded or that clearcut decisions cannot be reached using the basic survey data,
(c) special situations exist, such as starting up a new process, following substantial changes to an existing process, when an OEL has been set for the first time, when there are unusual, infrequent or intermittent processes or operations, or
(d) the need for a range-finding exercise as a precursor to an epidemiological study.

A detailed survey has a precise objective; usually, it is to obtain accurate measurements of personal exposures averaged over appropriate periods. However, static measurements or a mixture of personal and static measurements may be made when it is necessary to identify specific sources of exposure or to delineate the nature and extent of a problem. The survey strategy will be particularly influenced by:

(a) the requirement for personal sampling techniques when data is required for comparison with OELs,
(b) choosing the appropriate sampling period if results are to be compared with OELs (which are defined over specific reference periods),
(c) the need to minimize errors which may be more critical than in a worst-case situation, and
(d) the need to obtain measures of average exposures (e.g. for epidemiological studies of substances presenting chronic effects) or peak exposure measurements (e.g. for acute toxicants).

The possibility of detailed statistical analysis of the data should be considered and used, where necessary, to guide the strategy.

The need to obtain accurate data suggests that, wherever possible and relevant, the entire period of an individual's exposure within a working day should be covered by one or more samples taken consecutively or periodically. If periodic measurements are undertaken, consideration needs to be given as to whether these should be fully randomized or stratified (see section 7 below).

There will be a need to consider temporal variations in exposure, within a shift, between shifts on the same or different days together with any potential seasonal variations. Variations between shifts may be very high especially, for example, where the time base for a process is different from that for the shift system.

Where the number of employees is small, it may be possible (and sensible) to sample all potentially exposed employees. The number of shifts, workdays or weeks to be covered will still need to be considered as indicated above. Where large numbers of employees are involved, sampling will need to be selective and a statistical approach used to select individuals for inclusion in the survey. Random sampling of the workforce can generate large numbers of samples. This can be controlled, to some extent, by dividing employees into homogeneous exposure groups. When developing a group sampling protocol, it must be recognised that there may be individuals who have unique exposure patterns and who cannot be assigned to a group. These individuals must be monitored separately. Sampling of a group should be performed on the basis of random selection of individuals. However, it is important to remember that, morally and, in most countries (including the UK), legally, employers have a duty to every employee. Hence, both individual results and group means are important. Once data become available, examination of them may suggest that a group is not homogeneous and may need refining. However, consistently high exposures for an individual or individuals are of more practical importance in protecting those individuals from potential ill-effects than whether they should be included in a particular exposure group.

Sufficient measurements will be required to ensure that the range of activities and exposures within the group are covered, including periods of peak exposures, where necessary. It is important to take account of operations which may occur at the beginning or end of a workperiod and which may significantly affect the overall exposure, e.g. weighing operations or machine cleaning. Many maintenance or cleaning operations are performed outside normal working hours and any such work should be considered for inclusion in a survey.

2.6 Routine surveys
Routine surveys involve periodic sampling of employees or selected static sites to meet well-defined, long-term objectives, which may include:

(a) checking that control measures remain adequate,
(b) following trends in exposure levels and patterns and ensuring continued compliance with OELs,

(c) providing data for epidemiological studies, and
(d) complying with national legislation.

Monitoring strategies designed to detect what may be relatively small rates of change in exposure patterns require careful design and control and may need relatively large numbers of measurements. Before implementing such a programme, it is sensible to consider the relative costs of monitoring or improving the control measures. The latter can eliminate the need for extensive, continuing sampling.

Selection of employees for sampling can be done on a similar basis to that for a detailed survey. In addition, the frequency with which the surveys are to be made needs to be considered. Some standards include procedures for determining the need for, and periodicity of, monitoring [5]. These usually depend on relating the frequency to exposure level expressed as fraction of the OEL. A mechanistic approach of this nature tends to ignore the quality and quantity of the available data. Whatever approach is used, the following factors have special significance:

(a) *The proximity of measured exposures to the OEL.* The closer to the OEL, the more likely non-compliance becomes, so that minor changes in conditions can result in over-exposure and the more frequent monitoring should be.
(b) *The effectiveness of controls.* Robust controls, giving stable operating conditions, need little routine monitoring.
(c) *The consequences of control failure (and the time to re-establish control if failure occurs).* The physical form of the materials being handled and their toxicity are important. Control failure can give rise to serious acute risks. Similarly if a substance has a steep dose–response curve, employees can be quickly placed at significant risk.
(d) *The nature of the process cycles.* A routine survey may need to ensure that the cyclic nature of a process is adequately covered, e.g. to ensure that periods of high or low exposure are measured to give a true overall picture of exposure. At the same time, the periodicity of the survey needs to be different from that of the process; otherwise, a biased picture of conditions can result. For example, sampling the same shift again and again at the same time of year can bias results where ventilation conditions change with the time of year and time of day.
(e) *The variability of exposure.* Where exposure in a group is highly variable, a greater frequency of monitoring may be necessary.

Consideration of all these factors can lead to a wide variation in the need for routine surveys and their periodicity. Intervals between periodic measurements may vary widely from less than a week to more than a year.

All types of errors need to be minimized and capable of estimation, otherwise difficulties will be experienced in recognizing trends. In the long term, changes in sampling/analytical methods may occur for various reasons. These changes will need to be documented and the relationship between results from the two methods established. In a long survey, the personnel carrying out the monitoring will change so that data recording needs to be of a high standard. Sufficient details of processes, equipment, tasks, controls, etc., (and any changes) need to be recorded if new staff are to spot changes and be able to account for these changes.

2.7 Quality

The quality of a survey must match its purpose. There are two aspects to this:

(a) method quality, and
(b) strategy quality.

Method quality relates to measurement techniques which must be validated to an appropriate standard, commensurate with the survey objectives. Guidance on some aspects of method quality is given in MDHS 71 [9]. In some cases, it may be appropriate for laboratories to be involved in performance assessment schemes, such as the Workplace Analysis Scheme for Proficiency (WASP), Regular Inter-laboratory Counting Exchange (RICE) Scheme for asbestos (administered by the HSE), or alternative, international or industry/company schemes.

Accreditation of a laboratory by the National Measurement Accreditation Scheme (NAMAS) may also be a relevant indicator of a laboratory's performance capabilities. A CEN standard [10] is under development for performance standards of measurement methods.

Strategy quality is more difficult to define but it relates to the ability to meet the survey objectives in a structured, cost-effective and efficient manner.

The quality of any product is related to how well it meets customers' requirements and to the quality of the processes which are used to produce it. Within the UK, BS 5750 addresses such considerations. This type of approach is as valid in setting up and operating an occupational hygiene survey as it is to any other product. The procedures by which workplace measurement programmes are established and managed, samples obtained and analysed and data interpreted can all be incorporated into quality systems.

In formulating any programme, quality principles can be built in at all stages. Measurement variations can be assessed using blank, spiked or side-by-side samples, for example. Where appropriate, survey protocols, procedures and objectives should be documented. Reported data, information and any conclusions should be sufficient to allow others to arrive at similar conclusions.

The quality of the whole operation ultimately depends on the personnel involved and their knowledge, experience and skills.

One aspect which is often overlooked is the co-operation, active commitment and understanding of management, supervisors and workforce at all stages in the process. An effective exchange of information is essential if a survey is to be carried out in an atmosphere of co-operation and trust. The person undertaking the survey needs the best available information from management and the workforce. Equally it is they who will have to bear the brunt of potential short-term disruptions to work during the survey and any longer-term costs and changes. With a real lack of empathy, positive sabotage is a possibility, together with the dismissal of any results and recommendations.

2.8 Sampling and analytical methods

In principle, the sampling and analytical methods chosen should fit the sampling strategy and not vice versa. However, on occasions, practicalities will dictate otherwise. If, for example, it is desirable to obtain a full-shift, 8-hour sample, this may not be possible if such a sampling period can result in breakthrough on the collection medium. Two or more samples may be required, increasing the resources which need to be committed to the survey. Equally, the limits of quantification of the analytical method may constrain the measurement time period, especially when trying to measure short-term exposures over, say, 10–15 minutes.

The acceptable level of validation of any measurement method will, to some extent, depend upon the intended use for the technique. A method employed in a basic survey may be acceptable with a lower level of validation compared with that for a routine survey.

For analytical methods, specificity, sensitivity, accuracy and precision, recovery efficiency, sample stability in storage and transport loss may all need to be considered.

If there is an established OEL, for compliance testing, one rule of thumb is that the limit of quantification of the analytical method should ideally be below one-tenth of the exposure limit for a sample collected over the time period assigned to the limit value.

For the sampling method, the sampling device and collection medium must be suitable for the contaminant being collected and the analytical procedure that is available. For aerosols, the particle size range to be collected must be considered. For mists and liquid aerosols, consideration must be given to whether a mixture of liquid and vapour is to be collected or just liquid. For gases and vapours, preferential adsorption of other contaminants (in the workplace atmosphere) may occur on the chosen collection medium. With all contaminants, the capacity of the collection medium should be sufficient to cope with the likely loading over the intended sampling period. Sources of potential sample loss, e.g., side-wall deposition in some sampling devices, need to be recognised. For some substances, the choice of sampling or analytical method may be prescribed, e.g., for asbestos and lead.

2.9 Exposure data recording

As noted previously, the information and data recorded and reported must be sufficient to allow others to interpret the available material and arrive at the same or similar conclusions to a report's author. To this end, proposals have been made for core data recording requirements [11].

3. MIXTURES

It is probably rare for an employee to be exposed to a single substance, although one substance may dominate the exposure. However, the broad aims of sampling are the same whether exposure is to a single substance or a mixture and the approach to designing a suitable strategy remains the same.

Exposure to a number of substances may occur simultaneously (e.g. paint and adhesive solvents are frequently mixtures) or consecutively. Consecutive exposure may occur when an employee moves between jobs or where different parts of a process use different substances.

Exposure may occur over part of a shift, a whole shift, or longer. The significance of the pattern of exposure will depend on how the body handles the various substances. Long biological half-lives will lead to bio-accumulation and retention and increase the risk of interaction between substances. The pattern of mixed exposure may, therefore, affect the measurement strategy, the sampling and analytical methods and the interpretation of the results.

Some complex mixtures, especially process fumes, e.g. rubber fume, welding fume, have OELs in their own right. For other mixtures, various theoretical approaches have been developed to estimate the combined effects of simultaneous exposure to a number of substances. These have been extrapolated into schemes to check compliance with OELs in mixed exposure situations [12–14]. The detailed application of these schemes lies beyond the scope of this chapter and reference should be made to the original texts.

Four categories of joint toxic action have been identified:

(a) **Independent action**: Each component acts individually in the body in a way which is different from, and unaffected by, the effects of other components.
(b) **Additive action**: The combined toxic effects are the sum of the toxic effects of each component acting alone.
(c) **Synergistic action**: The combined toxic effects are greater than the sum of those of the components. (Potentiation is a similar effect where one component without a particular toxicity causes the toxic effect of another component to increase.)
(d) **Antagonistic action**: The combined toxic effects are less than the sum of the effects of each component acting alone.

As detailed information on the interactive toxicity of most substances is not available, it is generally unwise to assume independent action unless there is good evidence. Provided that substances are not acting synergistically or antagonistically, they are often assumed to act additively in the absence of information which indicates independent action.

The application of the so-called additive equation allows an estimation to be made of what may constitute adequate control of exposure when:

$$\frac{C_1}{L_1} + \frac{C_2}{L_2} + \frac{C_3}{L_3} + \ldots + \frac{C_n}{L_n} < 1$$

where C is the concentration of a given substance in the mixture which has an OEL of L, each being expressed in the same units. The equation cannot be readily applied to a mixture of substances which have OELs of different types, e.g., Occupational Exposure Standards (OESs) and Maximum Exposure Limits (MELs). OESs and MELs have different philosophical bases and different interpretations of compliance in the COSHH Regulations. An approach to the resolution of this problem has been suggested [see 14].

Even when substances have an effect on the same organ, their OELs may have been set on the basis of different lead toxic effects. For example, even in a homologous series of organic chemicals, some may have had OELs set on, say, liver damage and others on irritation. It is essential that the documentation for an OEL is consulted and the lead toxic effect identified before any OEL is used in the additive equation. It may be necessary to seek expert advice before considering any strategy for assessing exposure to mixtures.

When considering what to measure in a mixture, the following options can be considered:

(a) measure all or many of the components,
(b) measure a single substance or limited range of substances as a guide to control, or
(c) measure a surrogate material as a guide to exposure.

These options have been discussed in detail [see 15]. When compliance with OELs is to be considered, the individual components will need to be measured.

Quantification of the total 'mixture' is appropriate when an OEL is set for the mixture (e.g. white spirit, welding fume, rubber dust and fume). Care needs to be exercised because it may still be necessary to ensure that OELs for components are being complied with, e.g. nickel or chromium VI when welding stainless steel. Measuring total mixtures of dust for low toxicity dusts when no OEL is available may be appropriate. In the UK, total concentrations of any dust of 10 mg/m^3 (8-hour time-weighted average (8h-TWA)) or more would be regarded as a 'substantial' concentration under the COSHH Regulations.

It may be possible to identify one substance in a mixture which can be used as a marker such that control of that substance below its OEL will automatically ensure that the mixture is under control. There is a range of options in applying this approach or the use of surrogates [15].

4. DATA DISTRIBUTION MODELS

For decision making on various matters relating to the management and control of a workplace environment, it would be useful to have a mathematical model to describe the random variability of time-averaged contaminant concentrations. The period over which the data are to be assessed may be from one shift to beyond one year. Concern about the variability only arises when the measurement period is short when compared with the assessment period. A mathematical model would make it possible to make statements about the probability of occurrence of any given concentration value or about the probability of observing concentrations above or below some criterion value, e.g. OEL or action level.

The random distribution of many, but not all, sets of occupational hygiene data have a positive skew. The probability density function, which has attracted most attention, is the log-normal distribution. (See Fig. 1). This is the basis of most published work on the development of sampling strategies and processing of occupational

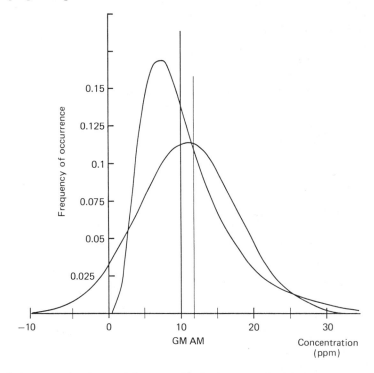

Fig. 1. Log-normal and normal frequency distribution curves having the same arithmetic mean. AM, arithmetic mean; GM, geometric mean.

hygiene data. The distribution of the data is described by two parameters, the mean and variance.

Not all data reported in the literature necessarily fit the log-normal model. Other skewed distributions or, alternatively, the normal distribution may be appropriate in some cases. Before relying on a particular model, the goodness-of-fit of the data should be tested, e.g. using a chi-square or Kolmogorov–Smirnov test.

For those with little knowledge of statistics, a more convenient method for evaluating the data against a particular statistical model is to use a probability plotting technique. The cumulative frequencies of normally distributed data produce a straight line when plotted on probability graph paper. A similar procedure can be used for log-normal data using log-normal probability paper. The method can be applied to small sample sizes and, in theory, as few as three results can be used (10-20 results are preferable). Application of the method is described, in greater detail, in several publications [2,5,24] Fig. 2 represents the results for a group of antibiotic workers. The slope of the line indicates the variability of the results. A steep slope indicates highly variable data and high standard deviations. An estimate of the geometric mean (GM) value is the 50% probability value which can be read directly from the graph. An estimate of the geometric standard deviation (GSD) can be calculated from the ratio:

Sec. 4] Data distribution models 187

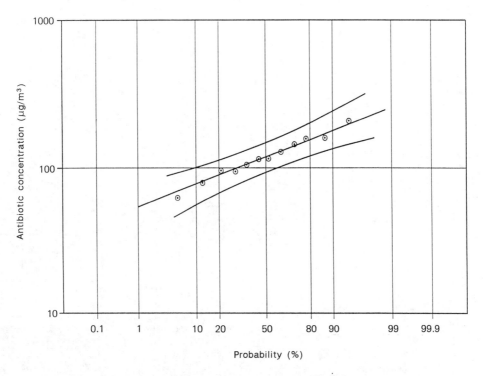

Fig. 2. Log probability plot of individual shift average antibiotic concentrations.

$$\text{GSD} = \frac{84\% \text{ value}}{50\% \text{ value}} = \frac{50\% \text{ value}}{16\% \text{ value}}$$

However, having plotted the individual results, provided the points lie close to a straight line rather than drawing the line by eye, it is better to take logarithms of the data and calculate the geometric mean and geometric standard deviation. A plot of the points GM × 1.65 × GSD (95th percentile) and GM/1.65 × GSD (5th percentile) allows the line to be drawn. The line should pass through the geometric mean at the 50th percentile for log-normally distributed data.

Plotting the data on graph paper and visually fitting a straight line or using the 5th and 95th percentiles is not a goodness-of-fit test. This can only be done by calculation. Application of a suitable method is discussed by Dewell [16]. Once the most appropriate distribution model has been identified, a regression equation for the data can be established and the probability plot constructed.

If the data are plotted directly onto log-normal probability paper and a straight-line plot is not obtained, the data may not be log-normally distributed. Fig. 3 shows a line which indicates that the data may come from two separate data sets, i.e. two different exposure groups (see section 5 below). Fig. 4 shows a curve which indicates that the data may not be log-normally distributed or, alternatively, the measuring device is saturating. Similarly, a curve, as shown in Fig. 5, may indicate non-

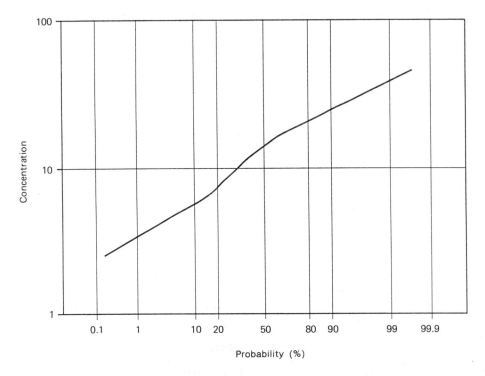

Fig. 3. Probability plot of a mixture of two distributions.

log-normality or that low exposures are not being accurately measured due to poor analytical sensitivity.

There are two significant consequences of having a skewed log-normal distribution for occupational hygiene data. Firstly, the number of results below the arithmetic mean is greater than the number above, for a set of experimental data. This means that, with a small data set, the estimated arithmetic mean is likely to be below the true value. Only if a large number of measurements are obtained would the estimated mean approach the true mean. This is important because it is the arithmetic mean of the airborne concentration which is related to the human body's accumulated uptake of a substance, a key factor in the development of long-term health effects.

Secondly, the larger the number of measurements made, the greater is the chance of obtaining a result in the upper tail of the distribution curve, possibly above the OEL. This could be interpreted as a failure to comply with the OEL. It could, therefore, be argued that, by minimizing the number of measurements, it will be easier to demonstrate compliance with an OEL. However, this should be resisted. It is better to obtain sufficient data to allow its interpretation to be put on a sound basis than to under-sample and bask in a false sense of security whilst having employees whose health may still be potentially at risk.

It is important to remember that the log-normal (or any other) statistical model can only be applied to measurements which are subject to independent sources of

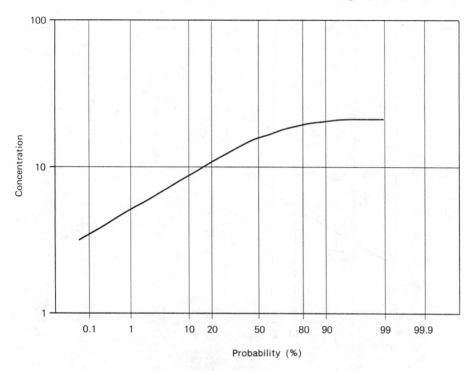

Fig. 4. Probability plot of right-truncated distribution.

random variation about the mean. For example, if an employee's work pattern is the same, day after day, the 8h-TWA samples obtained from successive shifts will probably exhibit log-normality. However, if the worker undertakes different tasks on successive shifts, there is a systematic variation introduced into the sampling results which are, therefore, unlikely to be log-normal. Extreme cases of such systematic changes may be seen in many batch process operations, e.g. in the fine chemical or pharmaceutical industries. Here, not only do feedstocks and products vary throughout the workplace, but individual substances are handled or produced on an intermittent, campaign basis with process conditions also subject to variation for any given product. A sampling strategy for such often highly variable conditions does not readily lend itself to a highly structured, rigid statistical approach to sampling over long time periods.

5. BIOLOGICAL CONSIDERATIONS

Assessment of exposures for epidemiological studies may be significantly different from that used for compliance assessment or assessment of workplace controls. A primary objective is to link tissue dose of a toxic agent (not just exposure) to effect.

The pharmacological or toxological processes linking exposures to adverse health effects, therefore, need to be taken into account when determining an appropriate sampling strategy. This is also true if a survey is to involve both biological and

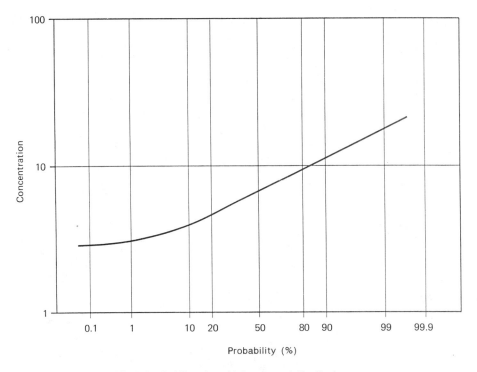

Fig. 5. Probability plot of left-truncated distribution.

airborne contaminant sampling. The relationship between initial exposure and expression in biological media (e.g. blood, urine or exhaled breath) needs to be understood if the measurements of airborne contaminant concentrations are to be linked with concentrations of the substance, or its metabolites, in biological samples. The toxicokinetics and pharmacokinetics may, therefore, have to be taken into account when selecting the measurement averaging time and period of interest.

In practice, there is no single kinetic model for a dose–response relationship. There is also a lack of suitable human toxicokinetic information for a large number of hazardous substances, with limited knowledge on damage repair mechanisms and their kinetics. Therefore, it may often be necessary to obtain expert advice prior to establishing a sampling programme where the results are to be correlated with health surveillance programmes, especially if they involve biological or biological effect monitoring.

Long-term, chronic effects are generally considered to be proportional to the mean exposure. Theoretical models have been developed [17] which suggest that transient peak exposures or fluctuations are of no consequence in terms of tissue damage (beyond their contribution to the mean), when the biological half-life of the substance is greater than 40 hours. Tissue damage is then related to the mean exposure and time. Advantage could be taken of this fact to minimize the number of samples taken, e.g $1 \times 40h$-TWA sample, in place of $5 \times 8h$-TWA samples, provided the kinetics remain linear. This should normally be the case under occupational conditions provided there are no synergistic or antagonistic effects related to concurrent exposures to other agents (see section 3 above) or allergenic responses to sensitizers. However, when considering the sampling strategy, it should be remembered that long-term sampling will reduce the amount of information about specific parts of the process. Short-term measurements may be required for evaluation of process controls. Also, the vast majority of OELs relating to chronic toxicity are based on an 8-hour standard and regulatory bodies require appropriate samples to be taken to demonstrate compliance with these OELs.

For many acute toxicants or irritants, biological half-lives range from seconds to a few hours. Variability in exposure is likely to be rapidly and efficiently transmitted into tissue damage which may occur within the time-scale of a single shift. Many acute responses are non-linear and responses vary from the irreversible, e.g. pulmonary oedema, to the temporary, e.g. irritation.

Risk factors associated with processes can vary widely. A continuous process, with good standards of containment and little manual intervention, should show little variation in emissions to the workplace atmosphere and is readily characterized. Intermittent or batch processes, perhaps with significant manual intervention, are potentially highly variable and difficult to characterize. It is, therefore, difficult to develop a general strategy for evaluating short-term exposures. Where regulatory authorities set short-term limits, it is essential that sampling is undertaken to demonstrate compliance with these limits and to aid in the development of adequate control measures where these are not present.

However, an occupational hygiene programme relying on sample collection on a filter or adsorbent, followed by analysis, is not adequate for routine identification of excursions above the OEL when serious acute effects may result. Even if an excursion is detected, the event is past and the data primarily of historical interest.

For substances which may cause serious, potentially irreversible acute effects, continuous monitoring, using either self-contained instruments, multi-point sampling systems or multiplexed sensors, may be a more appropriate approach provided suitable instruments are available. Depending upon the system, exposure levels can be measured almost instantaneously and the system used to provide an alarm when high level excursions require the implementation of emergency or evacuation procedures. A similar strategy may be appropriate for monitoring substances with less serious, reversible effects, where, because of process variability and possible intermittent operations, normal occupational hygiene monitoring programmes are difficult to implement.

6. SELECTION OF SAMPLE POPULATIONS

Regardless of the survey objective (e.g. an industry-wide, prospective, epidemiological study over several years or a one-off exercise covering a few workers in a single company), any data gathering should be to a strategy which recognizes the inherent statistical nature of assessing exposure.

Sampling every employee with potential exposure to a particular contaminant is frequently not a viable proposition. Several approaches are available for selecting sample populations. It is an area which is still fairly contentious and each method has its strengths and limitations. Selection of the appropriate method will be governed by particular survey objectives and resources.

Where there is requirement to demonstrate compliance with a legal standard, regulatory organizations may promulgate their own mandatory sampling strategies. These may differ significantly from the approaches outlined below. Specific national or international compliance sampling strategies are not described here because of their potential variety and limited application. Reference should be made to the relevant legislation if such mandatory methods have to be used.

6.1 Prospective employee grouping

This concept has been developed for employee grouping as the basis of a stratified measurement strategy [18]. A homogeneous group was defined as one in which all persons had a similar exposure level and profile.

The value of grouping is that the variability of the sampling results should be lower for a well-defined group than for the exposed workforce as a whole and there is greater likelihood of being able to describe the variability by means of a theoretical model. The practical advantages are that resources can be concentrated on those groups at maximum risk or those with the highest exposure if the objective is to improve control.

As described [18], exposure zoning (employee grouping) involves the following steps:

(a) preparation of a chemical inventory for the facility and determination of chemical utilization by plant area,
(b) identification of chemicals in the inventory which are of particular concern because of their hazardous nature, and
(c) matching the chemicals of concern to plant areas and employees who come into contact with them. Employees are then allocated to exposure zones, each zone having the following criteria:

  (i) similarity of tasks (not necessarily exactly the same job),
  (ii) exposure to the same range of airborne contaminants (including by-products and intermediates),
  (iii) similarity of environment, i.e., process equipment, exposure sources and ventilation arrangements, and
  (iv) identifiability.

When a sufficient number of employees are sampled in each zone, the information obtained should adequately describe the exposure concentrations for all employees in the zone. Zones are not necessarily single, definable geographic areas. It was recognized that not all employees can be fitted into appropriate zones and some flexibility is needed to accommodate these individuals.

Criterion (ii) is potentially flexible to suit the survey objectives. If exposure to a single substance is the prime concern or, alternatively, all agents of interest are used in all areas, the criterion for agent similarity is automatically satisfied. When measuring exposure to multiple agents, and the substance use varies throughout a facility, the criterion for agent similarity should be satisfied by ensuring that all persons included in a zone are liable to have the same exposures to the same substances.

Criterion (iv) was a restriction originally placed on the concept to ensure that a worker was not allocated to more than one zone.

Successful zoning depends on a thorough knowledge of tasks, working techniques, agents, processes and personnel records.

When the group sizes in each zone are known, the number of workers to be sampled can be determined. Sufficient samples will normally be required to ensure that the range of exposures in the group is covered or can be defined with sufficient accuracy. This is discussed further in section 7 on sample sizes.

After completing a sampling exercise, the plant may be re-zoned for a subsequent survey. However, this should not be done retrospectively, based on the results. Retrospective zone adjustments may lead to violations of one of the four criteria outlined. Prospective re-zoning uses information from previous surveys to determine whether the number of zones needs to be changed. Results from previous surveys are useful for determining the zones where the bulk of the sampling effort should be directed in future surveys.

When analysing the results and the initial zone allocations, high results are usually of particular interest. It needs to be determined whether these are genuinely due to random variation in the results from a homogeneous group or whether they are due to non-random effects from a non-homogeneous grouping. An initial grouping may contain an initially unrecognized sub-group of individuals with consistently higher exposures than the rest of the group. A useful rule of thumb is that no individual's exposure should be less than half or greater than twice the group mean [1] (see also section 6.2). If so, the individual may require allocation to another group or separate treatment.

6.2 Retrospective employee grouping

It has been suggested [17] that random sampling of the total exposed population is the most satisfactory way of obtaining data for the assessment of long-term exposures, particularly for epidemiological purposes. Employees are then retrospectively placed in homogeneous exposure groups, based on the actual data obtained. The technique is beyond the scope of this book and is perhaps best employed in specialized areas such as epidemiology. If the objective is, for example, to undertake a health risk assessment or measure for compliance with an OEL, random sampling of the whole exposed population may not be necessary. Many process operations may be under extremely good control with low employee exposures so that measurement of airborne contaminated levels could be unnecessary.

However, it is useful to understand the concepts involved in retrospective grouping.

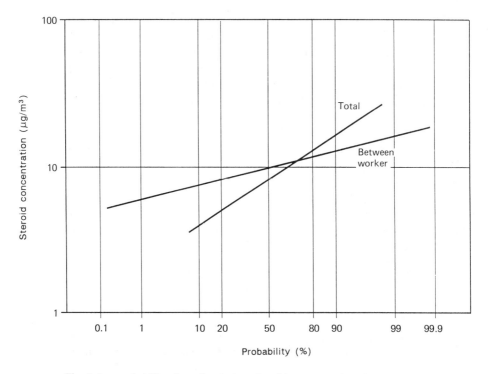

Fig. 6. Log-probability plots of typical total and between-worker distributions.

A homogeneous exposure group has been defined [17], based on the assumption that the exposure data are log-normally distributed. Data obtained from the population of employees sampled will have a geometric mean (GM) and geometric standard deviation (GSD). Each individual sampled in the population has a personal, day-to-day exposure distribution which will differ from that of the whole population. The individual workers' exposure distribution (within-worker) can be characterized by the geometric mean for the individual and the geometric standard deviation. The distribution of these individual means (between-worker distribution) should be log-normal with the same GM as the distribution for the total population.

The graph of Fig. 6 shows, by the difference in slopes, that the GSD for the total population is greater than that for the between-worker distribution. This is to be expected as the total population GSD represents both the within-worker and between-worker exposure variability. The distribution of individual means (the between-worker distribution) is used to define a monomorphic (homogenous) exposure group. This is arbitrarily defined as a group in which 95% of the individual mean exposures comprise a single log-normal distribution and lie within a factor of 2. This implies that the ratio of the 97.5 to 2.5 percentile is not greater than 2 and the geometric standard deviation for the between-worker variation would be less than 1.2.

It is suggested that a low factor for the ratio of the 97.5 to 2.5 percentiles indicates that the exposure variation is governed by the process and environmental conditions and a large factor indicates major contributions from individual tasks or working practices. This is an important point for consideration as it suggests that, where employee practice significantly contributes to exposure, observation is a poor way of assigning employees to groups by the prospective method. Given that initial walk-through surveys to assign groups cannot usually devote time to the observation of all individuals, this is to be expected so that increased care will be required when interpreting the data obtained. This is recognized in the HSE Guidance Note EH 42 [1] where it is recommended that, in any prospective grouping exercise, individual (mean) exposures should be within a range of one half to twice the group mean (a monomorphic group with 95% of individual means within a factor of 4 and a between-worker GSD of 1.4 or lower). If exposures are outside the range, it is suggested that jobs should be re-evaluated and employees assigned to more appropriate groups for future measurements.

7. SAMPLE POPULATION SIZE

Having selected a cohort for sampling, by whatever method, it is necessary to decide on the number of samples to be collected. This can be done on a statistical basis as described below, in which case, sample numbers can become extremely large and resources may not be adequate for such an extensive exercise. These large sample sizes result from the fact that workplace exposures are highly variable with respect to time and space. This contrasts with the much lower order of variation in the performance of the sampling or analytical methods which consequently require few measurements to be taken for method validation purposes.

It has been suggested that, from any given cohort, at least 1 in 10 are sampled [1, 19]. Corn [20] suggested that at least three samples should be taken before any statement of results is made and that additional samples should be taken if the results exceed a 25% spread. Such rules of thumb should be used with care as they can significantly affect data quality. In small populations, sampling the whole population should be considered to minimize any uncertainties.

One method of determining the number of measurements to make requires prior knowledge of an estimate of the mean and standard deviation. This can be obtained from a preliminary survey or previous experience in similar circumstances. Experience suggests that log-normally distributed airborne contaminant concentrations have geometric standard deviations between about 1.3 and 3.0. A figure of 2.0 is often selected to estimate the number of samples required to give an estimate of the population mean within any desired confidence interval, as follows:

$$\text{Number of samples } n = (t \cdot V/E)^2,$$

where t is read from a table of t-distribution values for some chosen level of confidence for infinite degrees of freedom, V is the coefficient of variation and E is a level of error (acceptable or chosen). The equation assumes that the sample population is infinite or approximately so [16].

If the relevant mean and standard deviation is not known and assumptions cannot be made, provided the data are normally or log-normally distributed, an alternative method of determining sample size is that of the National Institute for Occupational Safety and Health (NIOSH) [2]. Here, an initial decision is made that at least one result in the sample, to be taken from a population, should be in the top T% with C% confidence.

The population (cohort) size is known and presumed to be homogeneous, e.g. a group of workers doing the same job. For example, a homogeneous group of size $N = 30$ exists and it is required that at least one sampling result should be in the top 10% (i.e. one of the three highest results from 30) with 90% confidence.

Table 1. Size of partial sample for top 10% and 90% confidence

Size of group (N)	8	9	10	11–12	13–14	15–17	18–20	21–24	25–29	30–37	38–49	50
Number of samples required (n)	7	8	9	10	11	12	13	14	15	16	17	18	22

From Table 1, the required number of samples n would be 16. Conversely, it should be remembered that there is a 10% probability of missing all workers in the top 10%.

Technical Appendix A of the NIOSH Occupational Exposure Sampling Strategy Manual [2] gives tables for alternative criteria (sample size for top 10% and 95% confidence; sample size for top 20% with 90 or 95% confidence).

An alternative method of selecting the number of samples to be taken is available when it is required to determine a given standard of compliance with an OEL, action level or other standard with a certain level of confidence. The idea that, for chronic effects, exposure should be evaluated in terms of the mean airborne contaminant concentration goes back to the early 1950s [21]. More recently, work has focused on statistical methods for testing the means of log-normally distributed exposures relative to limits [22]. An expression has been developed which links sample size requirements to test the arithmetic mean exposure (AM) against the OEL. Table 2 gives examples of the number of samples required to test compliance with an OEL at a 95% significance level and with 90% power [17].

Table 2. Sample size (n) requirements of the test of the mean exposure for 95% significance level and 90% power

AM/OEL	GSD				
	1.5	2	2.5	3	3.5
0.1	2	6	13	21	30
0.25	3	10	19	30	43
0.5	7	21	41	67	96
0.75	25	82	164	266	384

Since the mean value from a log-normal distribution is independent of the averaging time of the measurements, it is possible to evaluate the frequency with which a short-term exposure limit (STEL) may be exceeded based solely on a knowledge of the arithmetic mean (AM) [17]. This can potentially eliminate a requirement for additional short-term sampling, provided full-shift data are available. For example, if it can be demonstrated that the AM < STEL/4, no more than 5% of the short-term exposures can be expected to exceed the STEL regardless of the variance of the distribution.

Knowing the frequency with which a limit is exceeded may be of value in determining compliance with a short-term limit, particularly if exposure is essentially uniform throughout the day with only random variations around the mean.

Tasks or jobs which have systematic variations in exposure can still require short-term samples to be taken to identify parts of the work which consistently give rise to elevated short-term airborne concentrations. Knowing the excursion frequency does not necessarily give an indication of the airborne concentrations obtained during excursions. These data themselves may be important in determining whether a health risk exists or compliance with a STEL is being achieved.

Whichever method is used for selecting the size of the sample population, the selection of individuals for sampling should ideally be done using either random number tables or a computer or calculator with a random number program.

8. MEASUREMENT STRATEGIES

8.1 Measurement patterns

The measurement pattern may be influenced by a number of practical issues as indicated above. These can include:

(a) sampling and analytical resources available,
(b) availability of staff to take samples,
(c) location of employees and work operations,
(d) occupational exposure variation (intra-day and inter-day),
(e) precision and accuracy of sampling and analytical methods,

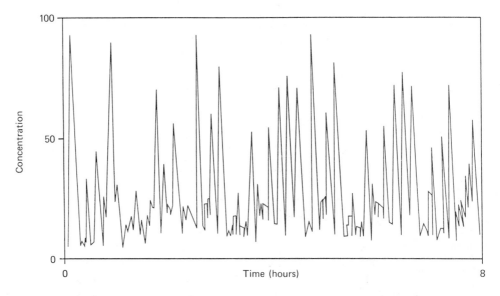

Fig. 7. Typical exposure pattern with random variations around a mean exposure.

(f) number of samples needed to attain the required accuracy for the exposure measurement, and

(g) type of measurements required, i.e., short-term, 8h-TWA, long-term, etc., or combinations thereof.

Figs. 7 and 8 illustrate some possible exposure patterns, based on a normal '8 hour' workday. Sampling patterns need to take account of exposure patterns if representative data are to be obtained. Possible sampling patterns are outlined in Fig. 9 for determination of an 8-hour average exposure.

Full-period single samples
These normally cover the full period of the relevant standard, e.g., 8 hours for an 8h-TWA (or 10–15 minutes for a STEL). The samples provide the second-best estimate of the average exposure over the period of interest.

Full-period consecutive samples
One or several samples of equal or unequal timespan are obtained during the entire period of the appropriate standard, e.g. 4×2 hour or $1 \times 6 + 2$ hour for an 8h-TWA. This methodology provides the 'best' results in that it gives the narrowest confidence limits on the exposure measurement during the period. The sampling periods can be chosen to cover task changes and thereby provide some data relating to the variation in exposure levels during the total period sampled.

Partial-period consecutive samples
One or several samples of equal or unequal duration are obtained covering only a portion of the period of interest. This approach is, perhaps, third in preference. The

Sec. 8]   Measurement strategies   199

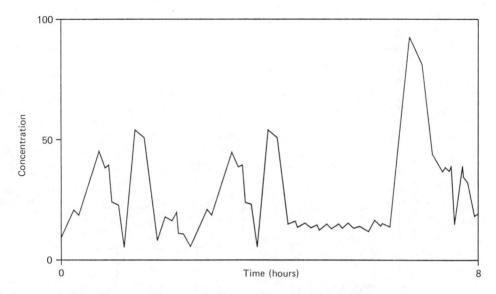

Fig. 8. Typical exposure pattern with random and systematic variations in exposure.

major problem is how to handle the unsampled periods. Professional judgement may allow inferences to be made about the unsampled periods provided reliable information is available on the tasks or process. NIOSH [2] recommend that at least 70–80% of the full period is sampled.

Grab samples

Grab samples, usually lasting only a few minutes or seconds, are taken at random intervals over the period of interest. The minimum number required for a homogeneous exposure period may be established by statistical analysis. The number of samples required may be substantial and increases as the sample time decreases. As an alternative, Table 3 may be used as a guide [5].

Table 3. Minimum number of samples as a function of sample duration

Sample duration	Minimum number of samples per shift
10 seconds	30
1 minute	20
5 minutes	12
15 minutes	4
30 minutes	3
1 hour	2

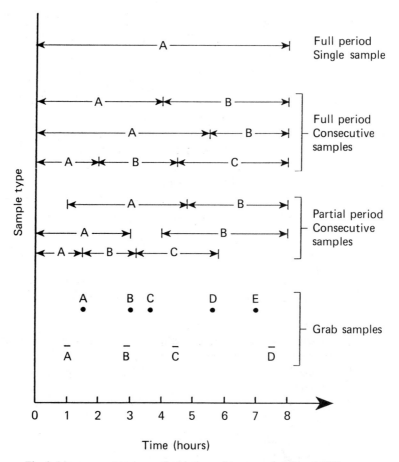

Fig. 9. Measurement types applicable to an 8-hour workshift and OEL.

Table 3 is based on the assumption that approximately 25% of the exposure duration should be sampled, provided that the working period does not contain significant systematic variations in exposure. With very short sample durations, this still requires a large number of samples, e.g. 720 for a 10-second sample duration. This is not practically feasible. Sufficient statistical stability is reached with 30 samples per shift as the greatest rate of change for any measure of dispersion, with respect to sample size, occurs when the sample size is smaller than 30.

Where systematic changes in exposure levels occur during a period of interest, a stratified random sampling exercise can make best use of the available resources. Stratified sampling consists of sharing the total number of measurements among the periods of differing exposure (each of which is fairly homogeneous) so that the number in each is proportional to the length of time involved, the measurements in each stratum being made in a random manner. Simple random sampling, under the same circumstances, can place undue emphasis on one phase or stratum, so that a source of sampling error would be introduced. The standard error of the sample would, therefore, be higher than that of the corresponding stratified sample of the same size.

Systematic sampling, i.e. making measurements at equal time intervals over the full assessment period, can be used where there is no systematic variation of contaminant concentration over that time.

Neither systematic nor stratified sampling methods are appropriate where there is relatively high-frequency cyclic variation in contaminant concentrations.

The above discussion also applies to the problem of estimating, for epidemiological purposes, the average exposure level of a group of employees over a much longer assessment period, by measuring 8-hour, time-averaged exposures of a selection of them for a number of shifts within that period. Timing of the sampling exercise can be random, stratified random or systematic. In practice, the last-mentioned approach is likely to be the most convenient to use. Due care is necessary, in this case, to ensure that sampling periods do not coincide with systematic, e.g. seasonal, changes in pollutant concentrations.

It may be convenient to plan long-term air sampling programmes on an annual basis. There is, thus, a finite population of N 8-hour, time-averaged exposures where N is the number of shifts during the year times the number of employees in the group to be studied. Frequency of sampling can be expressed as deciding what annual sample size, n, is sufficient to obtain good estimates of the parameters of the finite population, where

$$n = rk$$

where r is the number of persons selected for each trial and k is the number of trials per annum.

A pilot exercise may be required to obtain initial estimates of the mean and standard deviations. With these initial estimates, provided the sample population is large, e.g. the approximately 200 working days in a year, this can be assumed to be an infinite population and the number of samples

$$n = (t \cdot V/E)^2$$

as described in section 7.

In the case of monitoring for compliance with an exposure limit, it can be argued that the frequency of air sampling trials depends on how far employees' exposures differ from the OEL. This aspect is covered further in section 9.

8.2 Auto-correlation
The distribution of exposures received by a worker (or homogeneous group) can be described by the mean and variance. However, these parameters do not provide any information on the correlation of exposures measured at different times, i.e. the degree

to which a measurement result is influenced by the concentration which existed during the previous measurement. This is given by a third characteristic of the distribution, the auto-correlation function $p(h)$, which defines the relationship between air concentrations separated by h intervals of time, where h is referred to as the lag.

Auto-correlation is important because it affects the independence of samples which are collected during surveys. If a high degree of auto-correlation occurs, it may not be possible to draw valid inferences about the underlying population of exposures. Sufficient data is rarely collected in occupational hygiene surveys to allow auto-correlation functions to be analysed. The available evidence suggests that auto-correlation is not of great concern when full shift samples are obtained over several consecutive days.

Air-exchange rates have been used to estimate short-term auto-correlations which may exist in environments where mass transport of the contaminant is governed by turbulent diffusion [23]. Auto-correlation was low with high air change rates (10 air changes per hour) and high with low air change rates (1 air change per hour). If the air change rate is low, periods of hours between measurements may be required if unbiased estimates of short-term concentrations are to be obtained. Care should, therefore, be exercised when trying to fit data to theoretical models or when using these models to make predictions, particularly for short-term measurements obtained during a single shift, in a badly ventilated workplace.

8.3 Effects of averaging time.

As the averaging time of consecutive air samples covering the whole of an assessment period is increased, there is a smoothing of the variation in the measurements. Provided the conditions remain stable, the mean value is constant and independent of the averaging time, whereas the variance decreases with increasing averaging time. If the variance changes, the geometric mean also changes. The exposure distribution therefore changes with averaging time even when taken from the same environment, so that all samples must be collected over the same averaging period. However, minor variations are unlikely to have significant effects. As the precision of the estimated mean increases with increased averaging time, this would favour long-term sampling for the assessment of chronic health hazards. Equally, if an estimate of the time-averaged concentration for an assessment period of T hours is made by obtaining air samples of time, t, at random or systematic intervals (so that there are unsampled gaps within the assessment period), the precision of estimation improves, i.e. variance decreases as the size of the ratio T/t decreases towards 1.

However, smoothing of the variation in measured concentrations by increasing the averaging times may mean that valuable information is lost. For example, transient patterns of change, more persistent time trends in the workers' exposure or changes in process emission levels may be of considerable interest to the occupational hygienist or engineer concerned with evaluating the performance of control measures. The averaging time is, therefore, an important factor to be considered when deciding upon the overall strategy to be used for a particular survey or sampling programme.

9. RESULTS, INTERPRETATION AND ACTIONS

The interpretation of any set of measurements depends on the purpose of the exercise. The following section illustrates some approaches to the determination of compliance with OELs and the follow-up actions which may be required, particularly for the development of routine monitoring programmes. The section is not exhaustive in its approach and it is also necessary to recognize that 'compliance testing' still remains a contentious area.

A basic knowledge of statistics will be required for some of the methods described. Details of the methodology are not given but references are given for access to the necessary information.

9.1 General criteria for compliance

The comparison of measured exposures with OELs is probably the most common objective of a survey and, in most countries, OELs have some form of legal status. In the UK, most limits are covered by the Control of Substances Hazardous to Health Regulations (COSHH) 1988, exceptions being lead and asbestos.

Two types of limit are defined in the COSHH Regulations, the Maximum Exposure Limit (MEL) and the Occupational Exposure Standard (OES). Compliance with these limits is defined in the Regulations as follows:

> MEL—'The control of exposure shall, so far as the inhalation of that substance is concerned, only be treated as being adequate if the level of exposure is reduced so far as is reasonably practicable and in any case below the Maximum Exposure Limit.'

Arguably, this requires that the exposure of all employees must always be below the MEL. However, as noted above, the exposure of groups of workers or individuals varies, possibly with some form of skewed statistical distribution. Consequently, provided sufficient measurements are taken, results in the upper tail of the distribution will occur which are far from the mean concentration and, possibly, well above the MEL. Official guidance on compliance states that, where the limit is expressed over an 8-hour day, it is sufficient to demonstrate that the MEL is 'not normally exceeded, i.e. that an occasional result above the MEL is without real significance and is not indicative of a failure to maintain control'. This statement recognises that there will be variation in the data even for a well-controlled process or operation and it only remains to determine what is 'an occasional result'.

For substances with an OES, 'the control of exposure shall, so far as the inhalation of that substance is concerned, be treated as being adequate if:

(a) the OES is not exceeded, or
(b) where that OES is exceeded, the employer identifies the reasons for the standard being exceeded and takes appropriate action'.

Clearly, what is required to achieve compliance with a MEL is different from that for an OES.

The COSHH Regulations also place an onus on employers to develop their own working limits or 'in-house' OELs. Determining compliance with such a limit may be done on a different basis from a MEL or OES and would need to reflect the definition of the limit.

Oldham and Roach [21] agree that the only suitable measure for correlating with disease is the long-term average exposure. The definition of the MEL, in particular, suggests that it is not the average exposure of a group of workers or an individual that is of concern but some upper quantile. This idea is now generally applied to official OELs in many countries. The following sections set out a range of possible approaches to compliance testing.

9.2 The simplistic approach

The simplest approach is to require all measurements to be less than the OEL. If there is a limited amount of data, obtained in a preliminary survey, it may be appropriate for an initial judgement on compliance and any necessary further actions (see section 2.4.). However, with increasing amounts of data, the probability of obtaining a result above the OEL also increases, despite the value of the mean exposure.

An approach to compliance which allows no excursions above an OEL, or only isolated excursions, is intrinsically weak in that it potentially discourages measurement and encourages decisions to be made with poor quality data.

In a proposed European Standard [5], two approaches to evaluating compliance with OELs are set out. The first requires all measured 8-hour, time-weighted average concentrations to be below the limit and all exposure peaks during a shift to be below the short-term exposure limit. The measured concentration (C) is divided by the OEL to produce a dimensionless index of exposure (I) which is applied as follows:

(a) For the first sample, $I < 0.1$, exposure is below the OEL and, if it is representative of long-term conditions, no further measurements are required.
(b) If $I < 0.25$ for three different shift measurements, exposure is below the OEL and, if the data are representative of long-term conditions, no further measurements are required.
(c) If $I < 1.0$ for three different shifts, and the geometric mean of all measurements is $I < 0.5$, exposure is below the limit value.
(d) If $I > 1$, exposure is above the limit value.

In other cases which do not fit the scheme, the procedure leads to no decision.

One positive aspect of the scheme is that it recognizes, to some degree, that it is not necessarily sufficient to assess only the data in the tail of the data distribution but that the mean is also important. Fig. 10 shows a set of log-normal curves which have the same 95 percentile but different means and variances. If long-term health risks are related to the mean exposure, the closer the mean to the OEL, the greater the potential risk to health. In the event of some failure in the control measures, there will also be a greater risk of exceeding the OEL, the higher the mean.

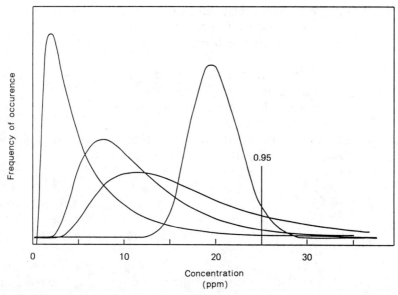

Fig. 10. A selection of log-normal probability curves having a common 95 percentile.

9.3 Estimated frequency of non-compliance

Table 1 illustrates the selection of sample sizes for ensuring that one result is in the top 10% with 90% confidence. The approach can be used for estimating compliance with short-term exposure limits during a single shift or compliance with 8-hour TWA OELs over a period of time, provided that the usual assumptions are made that variations in emissions, etc., are purely random and that there are no systematic variations in exposure. This is, in effect, a way of estimating the frequency of non-compliance. For individuals who are not versed in statistics, knowing that 'one result is in the top $T\%$ with $P\%$ confidence' is not helpful.

Application of a probability plot to estimating the frequency with which an OEL will be exceeded has much to commend it, at least in presentational terms. A graphical presentation is available and data can be reported in familiar concentration values. Details of the procedure are given in various sources [2,5,24]. It is applicable to any data set consisting of more than 10 measurements and, in practice, 20 measurements are usually enough to construct an acceptable probability plot. The data distribution model needs to be identified (log-normal; normal, etc.) and an initial plot of the data on log-normal and normal probability paper could be useful at this point. However, it is recommended that the data fit to the model should be tested. As noted in section 4, various test statistics are available. Correlation coefficients of normal and log-normal, ranked data can be calculated and their significance tested. The higher correlation coefficient indicates the more likely distribution. Using the chosen distribution, the regression equation is calculated using standard, least-square procedures. The procedure has been described [16]. The curve for data obtained during the manufacture of an antibiotic is shown in Fig. 2. Compliance with an OEL can be determined by reading directly from the regression line. The proposed CEN standard [5] suggests the following compliance criteria for the probability of exceeding the limit value:

(a) *Probability* $< 0.1\%$—compliance is achieved and no further measurement is necessary unless conditions change.
(b) *Probability* $> 0.1\%$; $< 5\%$—exposure seems to be below the limit value but periodic measurements are required to confirm the situation.
(c) *Probability* $> 5\%$—the situation is out of compliance and action should be taken.

The proposals in the standard are stringent although they are only for guidance. It is more usual to work to a probability of 5% as the indicator of compliance.

To take the test a step further, confidence limits can be calculated for another sample of data from the same population. Thus, if conditions remain the same, an estimate can be made of the concentrations which might have been obtained (with some level of confidence) had the measurements been made, for example, at different times during the day, different shifts or from different individuals in the group. The procedure is given [16]. When the confidence limits are plotted on probability paper, they produce an envelope similar to that shown in Fig. 2. If it is only required to determine the confidence limits at a given probability, e.g. 5%, these are readily calculated without the need to generate more data points to plot the envelope.

An alternative to probability plotting for estimating compliance is to carry out a one-sided tolerance test to answer the question; 'Are we 95 per cent certain that less than 5 per cent of all exposures received in the workplace are above the OEL'. The test can only be applied to normal distributions and log-normal data must first be log-transformed. The test is described analytically as:

$$T = Y + KS$$

For log-normally distributed data, T is the test statistic ($< \log$ OEL), Y is the mean of the log-transformed data, K is a single-sided tolerance factor (obtained from statistical tables) and S is the geometric standard deviation.

A major problem with one-sided tolerance tests is that the outcome is highly dependent on the number of results and the variance of the data. As occupational hygiene data sets frequently have few results and a high variance, the outcome from a single-sided tolerance test will often be one of 'no decision'.

A method exists for testing compliance based on the hypothesis that, if the mean exposure is above the OEL or some defined fraction (action level), the situation is out of compliance. As with the previous tests, the required number of measurements increases with increasing mean and variance. Table 2 illustrates this for the case where the significance of the test is at the 5% level and the power is 90%. An alternative graphical method is available for determining whether the mean of a set of measurements exceeds a limit value [22].

9.4 Measurements below the limit of detection

Statistical treatment of a data set is open to error when part of the data is unquantified because some of the results were below the limit of detection of the analytical method. The common method to compensate for this situation is to assign all the data points which are less than the limit of detection with a value which is one half of the detection limit. This produces a moderate bias, generally leading to underestimation of the geometric mean and geometric standard deviation but for most purposes the method is adequate.

9.5 Non-standard exposure patterns

Exposure limits relate to standard averaging times, the 8h-TWA values being based on the conventional 8-hour day, 5-day week work period. However, it is not uncommon for workers to put in overtime on top of the normal working period or, alternatively, have a different working week, e.g. 10-hour day, 4-day week. The principal reason for adjusting an exposure limit is that the longer working day results in a shorter recovery period between each exposure which can result in a greater accumulation of the substance in the body than might be expected. There are a number of complex pharmacokinetic models published for determining adjustment factors which can be applied to an exposure limit. However, their application is often limited by a paucity of data required for input into the models.

A simpler approach, which has some limitations, has been proposed [25]. An OEL reduction factor (RF) can be calculated from the formula

$$RF = 8/h \times (24 - h)/16$$

where h is the numbers of hours worked per day. The OEL is multiplied by the reduction factor to give the adjusted OEL and this figure is then compared with the adjusted OEL in the usual manner.

9.6 Statistical control charts

The use of control charts in manufacturing or analytical quality control is well established. In principle, the technique is applicable to occupational hygiene data to detect time trends in data from a routine survey and allow an evaluation to be made of changes in exposure or data variability. This requires rigorous control over all aspects of the strategy and measurement protocols. It is still relatively unusual for sufficient data of an adequate quality to be generated in occupational hygiene surveys. Consequently, the techniques are not described in detail here. An account has been given of the use and limitations of control charts for tracking changes in sample means and ranges against group means and action levels [26]. If properly set up, a 'cusum chart' [27,28] can give a better indication of underlying trends than a plot of sample means.

9.7 Periodic measurements

As discussed above, the frequency of routine measurements depends on a range of factors which relate to the characteristics of the substance and the process. The frequency also depends on the results of previous measurements. It will need to be increased if the exposure level approaches the OEL and decreased if the exposure level moves away from the OEL. This assumes that there are no significant changes in plant, process or materials in use which would warrant further sampling.

A proposed scheme for the regular monitoring of personal exposures is shown in Table 4. [23].

Table 4. The minimum time to be spent on regular monitoring of personal exposure

Man-shifts covered by sampling (per 10 employees)[a]	Personal exposure/OEL
1/month	1–2
1/quarter	0.5–1 or 2–4
1/annum	0.1–0.5 or 4–20
None	<0.1 or >20

[a] If <10 employees/shift, assume 10.

The scheme recognizes that, not only is routine monitoring of low exposure levels a waste of resources, but so is monitoring high levels. In the latter case, the primary objective is to improve workplace control measures to reduce exposures to acceptable levels.

In one scheme, proposed in a draft CEN standard [5], an initial time period (<1 week) is set depending on several factors, including:

(a) the work routine,
(b) the response time of the analytical laboratory and
(c) the type of OEL (STEL or 8h-TWA).

The basic measurement periodicity is then set at 8 time units. Depending on the results of previous measurements, the schedule is then modified, as shown in Table 5, by reference to four action levels:

$N1 = 0.4$ OEL
$N2 = 0.7$ OEL
$N3 = 1.0$ OEL
$N4 = 1.5$ OEL

The draft CEN standard [5] offers an alternative scheme as follows:

(a) The first periodic measurement is carried out within 16 weeks after the need for periodic measurements has been established. The maximum time interval to the next result depends on the result of the previous measurement.
(b) The interval is 64 weeks if the exposure concentration is $\leqslant 0.25$ OEL.
(c) The interval is 32 weeks if the exposure concentration is in the range >0.25 OEL $\leqslant 0.5$ OEL.
(d) The interval is 16 weeks if the exposure concentration >0.5 OEL $\leqslant$ OEL.

The choice of a 64-week baseline period in this scheme ensures that, over a period of time, the repeat measurements do not fall on the same week of the same month every year, reducing the risk of obtaining biased data due to seasonal effects.

Table 5. Decisions on modifications to monitoring schedules

Situation	Measurement result	Decision
1	$C \leqslant N1$ twice consecutively	Omit the following three measurements
2	$C \leqslant N2$	Continue basic schedule
3[a]	$N2 < C \leqslant N4$	A new measurement is taken during the next time unit
4[a]	$N2 < C \leqslant N4$ for two consecutive time units	An additional measurement is done in the four subsequent programmed intervals. If this interval is one time unit, immediate action should be taken to reduce exposure
5	$N3 < C < N4$ twice consecutively	Take immediate action to reduce exposure
6	$C > N4$	Immediate action to be taken to reduce exposure

[a] In situations 3 and 4, if $C > N3$, appropriate measures to improve control should be identified and implemented.

ACKNOWLEDGEMENT

The author wishes to record his appreciation to Dr R. J. Gardner (HSE), Mr J. W. Cherrie (University of Aberdeen) and Mr C. D. Money (Zeneca Specialties) who were the source of much of the original material [15] used as the basis for this chapter.

REFERENCES

[1] 'Monitoring Strategies for Toxic Substances', HSE Guidance Note EH 42, 1989, HMSO.
[2] 'Occupational Exposure Sampling Strategy Manual', N. A. Leidel, K. A. Busch and J. R. Lynch, DHEW (NIOSH) Publication No. 77–173, 1977.

[3] 'Technische Regeln für Gefährliche Arbeitsstoffe' (TRgA402) Messung und Beurteilung von Konzentrationengefährlicher Arbeitsstoffe (MAK), Bundesarbeitsblatt 11/1984, 55–61, Bundesminister für Arbeit und Sozialordnung, 1984.

[4] 'Guide d'Evaluation de l'Exposition au Risque Toxique sur les Lieux de Travail par Enchantillonage de l'Atmosphere', B. Herve-Bazin, Cahiers des Notes Documentaires, No 135, 265–285, 1989.

[5] 'Workplace Atmospheres: Guidance for the Assessment of Exposure to Chemical Agents for Comparison with Limit Values and Measurement Strategy,' PrEn 689, 1992, CEN.

[6] 'Occupational Exposure Limits and Monitoring Strategy', European Council of Chemical Manufacturers' Federations, 1984, CEFIC.

[7] 'Evaluation of Airborne Particles in the Work Environment', World Health Organization, Offset Publication No. 80, 1984, WHO.

[8] 'Hydrocarbon Distillate Vapour Composition: Prediction by Microcomputer', Technical Guide Series No. 2, 1982, British Occupational Hygiene Society.

[9] 'Analytical Quality in Workplace Air Monitoring', Methods for the Determination of Hazardous Substances, MDHS 71, HSE, 1991, HMSO.

[10] 'General Requirements for the Performance of Procedures for Workplace Measurements', PrEN 482, 1991, CEN.

[11] 'A Standard for the Presentation of Occupational Exposure Data', P. L. Beaumont and H. L. Dalrymple, *Ann. Occup. Hyg.* **36**, 79–98, 1992.

[12] 'TLVs, Threshold Limit Values for Chemical Substances in the Work Environment adopted by ACGIH', 1991–92, American Conference of Governmental Industrial Hygienists.

[13] 'Occupational Exposure Limits 1992', EH 40, HSE, 1992, HMSO.

[14] 'Guidance on Exposure Limits for Mixtures', 1993, Chemical Industries Association.

[15] 'Technical Guide No. 11: Sampling Strategies for Airborne Contaminants in the Workplace,' British Occupational Hygiene Society, 1993, H and H Scientific Consultants.

[16] 'Some Applications of Statistics in Occupational Hygiene: BOHS Technical Handbook Series No. 1,' P. Dewell, 1989, Science Reviews Ltd.

[17] 'Assessment of Long-Term Exposures to Toxic Substances in Air', S. M. Rappaport, *Ann. Occup. Hyg.* **35**, 61–121, 1991.

[18] 'Workplace Exposure Zones for Classification of Employee Exposures to Physical and Chemical Agents', M. Corn and N. A. Esmen, *Am. Ind. Hyg. Ass. J.* **40**, 47–57, 1979.

[19] 'Occupational Hygiene Measurement Strategy in the Pharmaceutical Industry', M. Rackham, G. V. McHattie and E. L. Teasdale, *J. Soc. Occup. Med.* **37**, 15–18, 1989.

[20] 'Strategies of Air Sampling', M. Corn, *Scand. J. Work Environ. Health.* **11**, 173–180 1985.

[21] 'A Sampling Procedure for Measuring Industrial Dust Exposure', P. Oldham and S. A. Roach, *Br. J. Ind. Med.*, **93**, 383–395, 1952.

[22] 'Die Schatzung des Zeitlichen Konzentrationsmittelwertes Gefährlicher Arbeit-

sstoffe in der Luft bei Stichprobenartigen Messungen,' W. Coenen and G. Riediger, *Staub-Reinhalt. Luft.* **38**, 402–409, 1978.
[23] 'A Most Rational Basis for Air Sampling Programs', S. A. Roach, *Ann. Occup. Hyg.* **20**, 65–84, 1977.
[24] 'Report No.4 : Papers Presented to the Institute 1983,' 1984 Institute of Occupational Hygienists.
[25] 'Occupational Exposure Limits for Novel Work Schedules,' R. S. Brief and R. A. Scala, *Am. Ind. Hyg. Ass. J.* **36**, 467–469, 1975.
[26] 'Statistical Control Charts: A Technique for Analysing Industrial Hygiene Data', N. C. Hawkins and B. D. Landenberger, *Appl. Occup. Environ. Hyg.*, **6**, 689–695, 1991.
[27] 'Guide to Process Control using Quality Control Chart Methods and Cusum Techniques', BS 5700: 1984, British Standards Institution.
[28] 'Guide to Data Analysis and Quality Control using Cusum Techniques' BS 5703: 1984, British Standards Institution.

12

Acute poisoning

Dr I. J. Lambert

SUMMARY

A significant number of incidents of chemical poisoning continue to occur in industry; over 800 serious incidents are reported each year, a number which is probably a significant under-estimate. When, after assessment, it is realized that hazards cannot be eliminated, employers should plan to minimize the effects of potential incidents, having regard to the characteristic properties of the toxic substances concerned. Few specific antidotes are available for the great majority of poisons; treatment relies on prompt and effective first aid, followed by hospital assessment and general supportive measures. This chapter discusses some problem areas in the management of cases of poisoning along with an overview of the facilities, training, equipment and information which should be provided.

1. INTRODUCTION

Poisoning is a major cause of illness and death; more than 4,000 people die each year in England and Wales from poisons absorbed either deliberately or accidentally. Over 100 000 cases of suspected poisoning are admitted to hospital each year in England and Wales, representing some 10 per cent of all acute admissions to hospital. Many of these poisonings result from deliberate self-administration but, of those which are accidental, most involve children under the age of five. Nevertheless, a significant number continue to occur in industry.

Table 1. Poisoning and gassing incidents reported in the UK

	1989–90	1990–91[a]
Fatal	20	23
Major, non-fatal	286	221
Causing more than 3 days absence from work	504	560
Total	810	804

[a] Provisional figures.
Source: Health and Safety Executive.

Figures on the latter are collected by the Health and Safety Executive. In 1990-91, provisional figures for the UK indicated 23 fatalities out of over 800 serious incidents (Table 1). However, these cases do not include minor incidents which can involve significant risks. There is also believed to be substantial under-reporting, with perhaps only a third of non-fatal incidents being notified.

Acute poisoning is defined in terms of the time which elapses between absorption of the poison and the appearance of symptoms, this usually being a matter of minutes or hours, rather than weeks. There is, however, no precise demarcation. Chronic 'loading' with a toxic compound may be tolerated by some people for months (or even years) without the appearance of overt symptoms until some relatively minor additional insult precipitates an episode of acute illness.

This can be seen, for example, in the ability of the body to cope with the anti-cholinesterase effects of organophosphorus compounds at low exposure levels without clinical symptoms, despite definite biochemical changes. However, when a critical cholinesterase level is reached, acute illness occurs.

Similarly, with prolonged exposures to lead, the body can accumulate a substantial burden of lead without noticeable effects until further exposures or metabolic changes precipitate overt lead poisoning.

Some substances can produce delayed acute effects; important examples of these are Paraquat (and the related bipyridylium herbicides) and the commonly-used drug, Paracetamol. Following ingestion of a single dose, severe, life-threatening symptoms may not appear for several days. The absence of early signs of poisoning can result in false reassurance to the uninformed and delay admission to hospital for immediate, much-needed treatment.

To manifest their toxic effects, poisons can enter the body by ingestion (i.e. via the mouth), by inhalation (into the lungs) or by direct absorption (through the skin). In industry, poisoning by inhalation is by far the most common route and the most important for the assessment of risks and establishment of adequate controls (see also Chapter 7).

2. INHALATION

Adverse effects can be produced by inhalation of the following classes of substances.

2.1 Asphyxiants
These are substances which interfere with the normal process of absorption of oxygen. Generally, they are either simple, inert gases (such as carbon dioxide or nitrogen) which take the place of oxygen in inspired air or chemicals which interfere with the effective uptake and use of oxygen by bodily tissues. A major example of the latter is carbon monoxide, a gas which combines with haemoglobin in the blood, thereby reducing its oxygen-carrying capacity and preventing the release of oxygen into body tissues.

2.2 Irritants
These substances, by contrast, cause damage by direct action on the respiratory tract and tissues of the lung. Examples include acid fumes and gases, such as sulphur dioxide, which cause immediate coughing, choking and subsequent pulmonary oedema (outpouring of fluid into the lungs). Other effects include tightness of the chest, shortness of breath, cough and frothy sputum, dizziness and collapse.

2.3 Narcotic agents
These agents depress the nervous system and cause weakness, dizziness, nausea and vomiting, blurring of vision and loss of co-ordination. Continued exposure can result in loss of consciousness and death. Such effects can occur with petroleum distillates, aromatic solvents and chlorinated hydrocarbons.

2.4 Other substances
Finally, a wide variety of miscellaneous poisons can be inhaled. Some of the more important examples, from an industrial point of view, include hydrogen sulphide, phosphine, arsine, metal fumes, formaldehyde, pesticides and herbicides.

As the incidence of solvent abuse (the deliberate inhalation of volatile compounds to produce mild-to-moderate narcosis) increases in society, so the availability of some of these substances in the industrial environment will inevitably lead to experimentation by some workers, with risk of adverse or even fatal results. A secondary risk arises from impairment of performance and increased likelihood of accidents. In industrial environments, where the more frequently abused substances are readily available, this potential problem must be assessed.

3. INGESTION
In contrast to inhalation, the risks of ingestion, in the industrial environment, are generally much lower. Where this does occur, poisons can exert effects by causing irritation or inflammation (usually by virtue of extremes of pH) or by absorption directly into the circulation.

The term 'secondary ingestion', is used where substances are inhaled into the lungs in particulate form, brought back into the throat (held in the protective mucous from the respiratory tract) and swallowed, with subsequent absorption from the stomach. This is thought to be relevant, for example, in the case of lead compounds.

4. SKIN CONTACT

The potential for large-scale acute absorption of toxic substances through the skin is generally limited. Direct skin effects can include irritation, redness, blistering or burns. However, certain substances are noted for their ability to be absorbed directly through the skin to produce systemic effects or even death. These include phenol, organophosphorus compounds, tetraethyl and tetramethyl lead, amino- and nitro-aromatic compounds.

5. RECOGNITION OF POISONING

In planning for the management of poisoning incidents, people often fail to appreciate that the basic diagnosis of poisoning can be difficult. On occasions, it may not be possible to decide whether a particular set of symptoms has been produced by poisoning or some other, unrelated cause. Often, the circumstances of the event (for example, a known escape of chemicals) can make the cause obvious. Less often, the occurrence of a specific and recognisable set of symptoms (either in one or several persons) can raise the suspicion that poisoning may have occurred. The diagnosis should, therefore, be considered within one of the following three very different scenarios:

(a) where an individual develops general symptoms which may or may not be due to a poison,
(b) where there has been exposure to an unknown substance (or group of substances), or
(c) where there has been exposure to a recognized poisonous substance.

These difficulties can be heightened with the realization that the toxic effects of many poisons can be non-specific—for example, collapse, abdominal pain or vomiting—and can be identical to the symptoms of a range of other illnesses. Furthermore, where there has been a chemical release of unknown dimensions, and where potentially exposed people have some knowledge of the early symptoms of exposure, it is easy for them to imagine the onset of poisoning. This phenomenon has been widely recognised in warfare when the use of chemical agents has been suspected. Even more dramatic have been symptoms exhibited as mass hysteria where large groups of people have become convinced that they have been exposed to some noxious agent and apparently all become ill. Such episodes can be extremely difficult to manage in the face of apparently conclusive evidence of poisoning.

It should also be borne in mind that life-threatening illnesses may occur for entirely unconnected reasons in people who are working with, or may have been exposed to, chemicals. The consequences can be potentially fatal if the true cause is overlooked. The need for proper evaluation of the situation—taking a history and making a full examination—is as important in suspected poisoning as in any other illness.

6. DOSE AND EFFECT

The quantity of any poison ingested can be difficult to estimate. The concentration of the agent in inhaled air in a gassing incident is often speculative. Even when ingestion has occurred and quantities can be more easily measured, calculation of the likely effects cannot usually be reliably made. For a limited range of substances, documented clinical experience from previous cases of poisoning can provide an estimate of the likely fatal dose but, for the majority of substances handled in industry, the only indications may come from animal experimentation. The most frequently quoted figure is the LD_{50}—the amount of the chemical which kills 50 per cent of a group of laboratory test animals, usually when administered orally; the dose is expressed in milligrams (or grams) per kilogram of animal body-weight. Owing to enormous differences in response between animal species (in terms of susceptibility to poisons), such figures should be interpreted with great caution and regarded as no more than an approximate indication of relative toxicity.

7. DIAGNOSIS

Acute poisoning is generally diagnosed from symptoms and signs rather than by laboratory or other investigative procedures. The latter can be more useful for confirmation and ongoing management, in severe cases. A significant interval usually occurs between the time of entry of a poison into the body and appearance of the first symptoms and signs; this is referred to as the latent stage. During the subsequent active stage, the effects of the poison become manifest. While some substances can produce characteristic reactions or groups of symptoms, most give rise to responses which are largely non-specific. Such general symptoms can include:

— weakness, fatigue and general malaise,
— nausea and vomiting,
— pain in the abdomen or limbs,
— drowsiness and impaired mental performance,
— diminishing consciousness,
— convulsions,
— shortness of breath or difficulty in breathing,
— prolonged unconsciousness and
— weak or rapid pulse.

In cases of severe poisoning, patients may survive initial effects only to experience complications subsequently. These can include damage to the liver or kidneys, infection or oedema of the lungs or failure of the heart and collapse of the circulation.

8. PRINCIPLES OF TREATMENT

The first aid treatment of poisoning should be directed at avoiding further ingestion or damage, maintaining circulation and respiration, arranging for further care and providing information which will assist in further management. All cases of poisoning, even those displaying no immediate ill-effects, should be properly evaluated; this normally involves admission to a hospital or suitable treatment centre. A summary of immediate, first aid measures which should be undertaken in cases of chemical poisoning is given in Table 2.

Table 2. First aid for chemical exposures

1. Protect yourself and the casualty. Remove from further exposure and use protective clothing and breathing apparatus, if indicated.
2. If the casualty is unconscious, ensure that the airway is clear and place in the recovery (semi-prone) position.
3. If the casualty has stopped breathing, give artificial respiration; if corrosive or highly toxic chemicals have been swallowed, use a proprietary respirator.
4. Check for a pulse and, if the heart has stopped, give external cardiac compression.
5. Remove any contaminated clothing and place in a sealed bag or in a safe, ventilated place.
6. If there is skin or eye contamination, wash with plenty of clean flowing water for at least 10 minutes. Cover skin burns with sterile dressings.
7. If chemicals have been ingested and the casualty is conscious, rinse the mouth with water and give one pint of water to drink immediately. If the chemical is corrosive, go on giving water (1/2 pint every 15 minutes) until further assistance is obtained.
   Do not attempt to induce vomiting.
   If the casualty is unconscious, do not attempt to give anything by mouth
8. If the casualty is confused, unco-ordinated or having fits, use only minimal restraint for safety.
9. Seek medical advice as soon as possible, even if the casualty appears well.

9. INDUCTION OF VOMITING

It is sometimes possible to avoid continued absorption of poisons by removal from the stomach. This is always better done by gastric lavage (i.e. stomach washout) as a hospital procedure but, in some circumstances, there may be a place for inducing vomiting while waiting for transfer to hospital. This should not be viewed, however, as a general first aid measure but must be carried out by trained personnel acting under standing instructions from a medical adviser. CIA has produced an advice card on this issue for medical personnel [1]. The decision to introduce this as a first aid measure depends on the nature of the chemical hazards and the ease of access to professional medical care.

There is no point in either inducing vomiting or washing out the stomach if the poison has already passed beyond the stomach. For most substances, this occurs about

4 hours after ingestion. Both procedures are also contra-indicated if the poison is corrosive, unless the risk of systemic toxicity is greater than that from further damage to tissues of the mouth and upper digestive tract. If a petroleum distillate has been swallowed, vomiting may result in aspiration into the lungs and subsequent pneumonia. If the patient is unconscious or semi-conscious, the only safe procedure is to carry out gastric lavage, in hospital, with full protection of the airway.

Where a decision has been made to induce vomiting, the preferred method is to use ipecacuanha, as 10–20 ml of tincture (BP) (for an adult) repeated after 20 minutes if there is no effect. Salt water, mustard, stimulating the pharynx and other methods are entirely contra-indicated.

10. ADSORBENTS

Activated charcoal is gaining recognition in the immediate management of certain types of chemical poisoning although acceptance as a first line of treatment in industrial practice in the UK is still uncommon. Administered by mouth (50 g for an adult), it prevents absorption of poisons from the stomach and hence, to be effective, must be given shortly after ingestion. It is particularly valuable against highly toxic agents, but the amount of charcoal must exceed the dose of the poison by a factor of 5 to 10 times by weight to provide adequate protection. It is of little use in the treatment of poisoning by corrosives, cyanide, alcohol, organic solvents and other water-insoluble substances.

The use of activated charcoal must be assessed in advance by qualified medical personnel and administered under standing orders by first aiders who have received specific training and instruction.

11. ANTIDOTES

It is a widely held belief that a large number of antidotes is available for specific treatment for many poisons. Unfortunately, only a small number of poisonous substances have direct antidotes and, even where these exist, side-effects and risks of use limit their application. The existence of an antidote does not warrant routine or automatic use; general and supportive measures remain the key to sound treatment.

Antidotes work in a variety of ways. They may combine directly with the poison to form an inactive product or they may compete with the poison for a receptor site. Other mechanisms include increasing the rate at which the poison is metabolized or retarding the conversion of the poison to a toxic metabolite. A list of most of the antidotes which have been found to be of value for chemical exposures in the work situation is given in Table 3. The use of such antidotes in treatment before admission to hospital needs to be evaluated by qualified medical staff, having regard to the particular nature of the risks and availability of medical and hospital care.

Table 3. Some specific antidotes useful in the treatment of industrial poisoning

Antidote	Preparation	Poison
Amyl nitrite	Breakable ampoules	Cyanides, Hydrogen sulphide
Atropine	Intravenous/intramuscular (i.v./i.m.) injection	Organophosphorus and carbamate insecticides
Calcium gluconate	Gel, injection, nebulizer solution	Hydrogen fluoride
Dicobalt edetate	i.v. injection	Cyanides
Dimercaprol	i.m. injection	Mercury, arsenic, antimony, thallium
Disulfiram	Oral	Nickel carbonyl
Ethanol	Oral/i.v. injection	Methanol, ethylene glycol
Methylene blue	i.v. injection	Amino/nitroaromatic compounds
Pralidoxime	i.v. injection	Organophosphorus insecticides
Penicillamine	Oral	Copper/lead
Calcium disodium edetate (EDTA)	i.v. infusion	Heavy metals, esp. lead
Sodium nitrite / Sodium thiosulphate (used in conjunction)	i.v. injection	Cyanides
Vitamin K_1	i.v. injection	Coumarin rodenticides

If specific treatments are used as a first aid measure, training in their use is required over and above that needed for the general qualification for first aid at work. Detailed reference to such training is given in the Approved Code of Practice to the First Aid at Work Regulations [2].

12. PROTECTION OF FIRST AIDERS

It is essential to consider, in advance, adequate provision for first aiders, rescue and ambulance personnel in the case of chemical exposure. If widespread exposure occurs, protective clothing and, if necessary, breathing apparatus will be needed. If the latter is not available, rescue should be left to professional rescue services. Where individuals are exposed, first aiders must be trained in the avoidance of risks and wear protective gloves and other clothing, as indicated. Contaminated clothing from casualties should be bagged and sealed and, most importantly, not allowed to be taken into ambulances or other confined spaces until this is done.

Mouth-to-mouth resuscitation may present particular risks if highly toxic or corrosive chemicals have been ingested. Protective devices, now widely available for use in resuscitation, provide adequate protection.

Inert gases are responsible each year for a number of totally unnecessary deaths where would-be rescuers enter confined spaces or underground systems to retrieve unconscious casualties only to be overcome themselves. No-one should attempt rescue in such situations unless he has been trained and is equipped with suitable breathing apparatus.

13. PROVIDING INFORMATION ON INCIDENTS

It is essential to try to communicate the precise circumstances of incidents to the medical staff involved in the continued care of casualties. First aiders, or people at the scene of incidents, should try, before evacuating casualties, to identify the individuals concerned, together with details of the substances which were being used at the time, the likely route(s) of exposure, the amounts of substance(s) involved, the duration of exposure and any protective equipment in use. Relevant medical conditions of casualties should be noted, if known: for example, epileptics on medication, asthmatics or sufferers from heart disease. This information should accompany casualties to hospital, along with details of the chemicals concerned, for example, by enclosing the packaging together with labels, data sheets, etc. or, failing that, a description of the physical characteristics or appearance of the substance(s).

14. PLANNING FOR ACUTE POISONING INCIDENTS

As part of the assessment of risks to health at work, all toxic substances on a site should be identified and appropriate controls instituted. Information and instruction about potential poisons at the workplace should be communicated to employees and to people (such as first aiders, occupational health nurses, medical advisers and others) who may be involved in responses to incidents. Data sheets or extracts of first aid advice should be readily available and regularly updated.

First aiders may need support from technical or laboratory staff in identifying particular hazards and appreciating the nature of the risks. First aid information on data sheets should not be accepted uncritically. When toxic risks are significant, it remains the responsibility of employers to confirm that the actions recommended are appropriate and complete; reference to other publications (e.g. [3], [4]) or professional advice may be required.

When particular hazards are present, manuals or cards should be prepared giving clear, unambiguous instructions to follow in an emergency. Examples of these can be found in the CIA's series of Chemical Exposure Treatment Cards [5], published as a booklet dealing with treatments for exposure to a large number of the commonly used gases and vapours. When a casualty needs to be transferred to hospital for further treatment, it should not be assumed that adequate information on specific treatments will be immediately available to the casualty department. Specific medical treatment advice may be prepared for the medical staff or others concerned. An example (for organophosphate poisoning) is given in Appendix 1 at the end of this chapter.

Responsible employers may wish to establish contact, in advance of any incidents, with the local casualty department or medical practice; visits by professional staff to the plant concerned can be of great benefit in establishing effective communication.

In the UK, poisons information services have been established and provide 24-hour emergency advice on all aspects of the management of specific poisoning incidents. Current telephone numbers for these services are given in Appendix 2.

15. FIRST AID PROVISION

In the UK, under the Health and Safety (First Aid) Regulations 1981, employers are required to provide first aid facilities which are adequate and appropriate for the circumstances. The associated Approved Code of Practice [2] gives details as to how these facilities should be realized. Although suitable contents for first aid boxes are suggested, the numbers of first aiders, their equipment and the establishment of a first aid room are seen to depend on the nature and extent of the risks to the workforce. The duration and frequency of training are prescribed but, as detailed above, when certain prescribed hazards are present, extra training, subject to the approval of the Health and Safety Executive, is required. Three specific hazards are currently prescribed: where there is a danger of poisoning by certain cyanides and related compounds, where there is a danger of burns from hydrofluoric acid and where there is a need for oxygen as an adjunct to resuscitation.

Paragraph 15 of the guidance to Regulation 3(2) goes on to state that 'In certain other cases (e.g. where there is a serious risk of gassing, or of exposure to toxic chemicals etc.), the first aider may need to undergo *such additional training as may be appropriate in the circumstances of the case*'.

Specialist advice is needed in most cases; where there is difficulty in arranging for specific training, advice should be obtained from the Employment Nursing Advisory Service.

16. RECORDS AND REPORTING

Employers should maintain adequate records of all incidents. If occupational health records on employees are kept, under the COSHH Regulations, arrangements should be made to link these to records of exposure. Under the Reporting of Injuries, Diseases and Dangerous Occurrences Regulations (RIDDOR) 1985, notification to the enforcing authorities is required for any accidents causing death, certain major injuries or incapacity for work in excess of 3 days. The category of major injuries includes 'acute illness plus any loss of consciousness resulting ... from the absorption of any substance by inhalation, ingestion or through the skin'. Certain other reports of occurrences are required; it is the responsibility of the employer—not the medical attendant or the individual—to make the necessary returns.

17. CONCLUSIONS

Avoidance of fatalities, following cases of chemical poisoning in industry, depends on the identification of hazards, provision of detailed information and adequate training for first aiders. Only in a few cases are specific antidotes appropriate and, where they are used, detailed supplementary training of first aiders is needed. Protection of first aiders and rescue staff should be considered and communication with local casualty departments—both before the occurrence of incidents and over details of specific medical treatments—should be undertaken.

REFERENCES

[1] 'Induction of Vomiting in the Treatment of Ingested Poisons', A4 laminated card, 1976, CIA.
[2] 'Health and Safety (First Aid) Regulations 1981', Approved Code of Practice, COP 42 (ISBN 0 11 885536 0), 1990, HMSO.
[3] 'Dangerous Chemicals Emergency First Aid Guide', Ed. Dr. A. Houston/Guys Poisons Unit, Croner Publications Ltd. 2nd edn, 1987 (revised periodically).
[4] 'Handbook of Poisoning', R. H. Dreisback and W. O. Robertson, 12th edn, 1987, Prentice Hall International. (ISBN 0 8385 3632 8)
[5] 'Chemical Exposure Treatment Cards', 2nd edn, 1991, Chemical Industries Association.

APPENDIX 1. EXAMPLE OF AN INFORMATION SHEET GIVING DETAILED MEDICAL ADVICE FOR MEDICAL STAFF ON ORGANOPHOSPHATE (OP) POISONING

Organophosphate compounds are cholinesterase inhibitors. The principal manifestations of poisoning are visual disturbances, respiratory difficulty and gastro-intestinal hyperactivity.

Atropine is the emergency antidote.

Initial symptoms and signs are headache, dizziness, nausea, muscular weakness, anxiety, anorexia, tremors of tongue and eyelids, miosis and blurred vision. These may be followed by tearing, salivation, abdominal cramps, diarrhoea, vomiting, sweating, slow pulse and muscular fasciculations, followed by muscular weakness. Eventually, pupils may become pinpoint and non-reactive and respiratory difficulty, pulmonary oedema, cyanosis, loss of sphincter control, loss of reflexes, convulsions, coma and heart block may occur.

In all cases of severe intoxication, inject as early as possible, preferably intravenously (IV), atropine sulphate, 2–4 mg for adults (0.04 to 0.08 mg/kg body-weight for children). Repeat every 3–10 minutes until adequately atropinized, as shown by dilated pupils, flushing of the skin and dry mouth. During the first hours of treatment, 20–80 mg or more of atropine sulphate may be required. If the intramuscular (IM) route has to be used to administer atropine sulphate, take care not to overdose when giving repeated injections, since distribution by the IM route takes longer than by the IV route.

While keeping the patient fully atropinised, administer also an oxime (cholinesterase reactivator) e.g., 2-PAM or P_2S (Contrathion) 1000-2000 mg IM or IV for adults or 250 mg for children. Repeat, if necessary, after 30 minutes with half the dose. Toxogonin is a more recent cholinesterase reactivator. It can be given instead of P_2S at a dose of 250 mg IM or IV for adults. The dose for children is 4–8 mg/kg. If necessary, repeat after 1–2 hours. Cholinesterase reactivators should be given within 36 hours after the onset of intoxication.

Convulsions can be treated with diazepam (e.g. Valium, Stesolid) 5–10 mg for adults (2–5 mg for children). Administration can be either i.v. or per rectum. Repeat if necessary until convulsions have ceased.

Contraindicated are morphine or other opiates, phenothiazines, succinylcholine, xanthine derivatives, epinephrine and barbiturates.

Keep airway clear and watch respiration. Artificial respiration may be required. Observe patient in hospital for at least 24 hours. Guard against dehydration and acidosis. If necessary, give intravenous infusion with dextrose or fructose solutions, electrolyte infusions, as indicated by haematocrit values, electrolyte spectrum, pH of the blood and diuresis.

The diagnosis of OP intoxication should be confirmed as soon as possible by determination of the cholinesterase activity in venous blood.

Reproduced from the 'Shell Agriculture Safety Guide', 1990, by kind permission of the Shell International Chemical Company Ltd.

APPENDIX 2. TELEPHONE NUMBERS OF UK POISONS CENTRES (AS AT JANUARY 1993)

Belfast:	0232-240503 Ext. 2140
Birmingham:	021-554 3801 Ext. 4109
Cardiff:	0222-709901
Dublin:	0103531-379966
Edinburgh:	031-229 2477 Ext. 4786
Leeds:	0532-430715 or 0532 432799 Ext. 3547
London:	071-635 9191
Newcastle:	091-232 1525 (9am-5pm) or 091-232 5131 (after 5pm)

Note. These numbers may be subject to change in the near future.

Poisons centres may be contacted 24 hours a day for advice on the specific management of individual cases. Advice is normally given only to medical personnel but, in an emergency, enquiries will be accepted from first aid or rescue workers.

13

Dermatitis

Dr P. C. Nicholson and Dr. A. Chojnacki

SUMMARY

This chapter covers the subject of occupational dermatitis, with particular reference to diseases caused by exposure to chemicals within the working environment. The scale of the problem is described with its impact on society and industry. The role of the skin as a major barrier to the ingress of substances into the body and its other functions as one of the body's major organs are also discussed.

Descriptions of skin diseases use terminology which can be confusing to the lay person. An attempt is made to de-mystify this terminology. Examples of skin diseases are given, with particular emphasis on chemically induced dermatitis. The major differences between irritant and allergic dermatitis are described. Predisposing and contributing factors to the development of chemically induced dermatitis are discussed with possible implications for pre-employment screening of potential employees.

Guidance is given regarding the basic concepts required to prevent occupational dermatitis occurring, including the use of toxicological studies to identify problem agents, workplace risk assessments, personal protective equipment, etc. The aims and types of health surveillance possible are discussed, as are investigations regarding potential, occupationally induced dermatitis.

Finally, this chapter discusses the prognosis of occupational dermatitis and the relative merits of rehabilitation and redeployment.

1. ARE OCCUPATIONAL SKIN DISEASES A PROBLEM TO INDUSTRY?

There can be no doubt that skin diseases caused by work, i.e., occupational dermatitis, is a major occupational health issue. It causes more days being lost due to ill-health than all other prescribed diseases put together. Indeed, statistics from the Department of Health, recorded between 1984–85, show that dermatological conditions accounted for 2.5% of all periods of absence of over 8 weeks' duration and that dermatitis accounted for 1% of all days lost due to sickness.

Not all these skin problems have an occupational cause but it is believed that 11% of all cases of hand dermatitis are due to occupation, resulting in a considerable number of cases. Indeed, in the USA, it is reported that dermatological disease occurs at a rate of 2.85 per 1000 full-time workers each year.

It is also appreciated in the USA, as in the UK, that reported cases are simply the tip of the iceberg and that skin disease is estimated to cost American industry around a billion dollars annually. Consequently, controlling occupational dermatitis could potentially save a great deal, both in monetary terms (through lost output) and in the reduction of personal disability and suffering.

2. THE ROLE AND FUNCTION OF THE SKIN

The skin should not be considered as a simple covering for the body but as one of the body's major organs. It is true that the chief function of the skin is that of protection and that this is of particular importance when considering possible exposure to chemical agents encountered in the workplace. However, it should be remembered that the skin has many other important functions. These include aiding temperature regulation, the excretion of waste products, acting as an organ of sensation and touch and protecting more deeply located tissues from ultra-violet radiation, harmful bacteria and physical agents.

In addition, the skin can produce the important vitamin D which protects against the disease, rickets. It is also important to realize that the skin is constantly being renewed, the thick outer layer (horny layer or stratum corneum) being replaced completely every one to three weeks. (See Fig. 1 below which shows, in simple terms, the basic structure of the skin, highlighting its division into layers.)

As mentioned above, the skin has an extremely important role in preventing harmful substances from entering the body. The main barrier to penetration is the horny layer, which is a thin layer of approximately 100 μm thickness (about the thickness of tissue paper). It is composed of dead, partly dessicated epidermal cells which contain concentrations of protein, known as keratin. This is a tough, hard substance. However, this barrier is breached by the presence of hair follicles and sweat ducts which provide channels for the passage of substances deposited on the skin surface to capillaries in the deep layers of the skin and, hence, into the general circulation.

The contribution of the sweat glands and hair follicles to the total permeability of the skin to most substances can be shown to be negligible (approximately 1%) but particles of very small size (e.g. <1 μm) may penetrate by these routes, by-passing the protective barrier. The horny layer itself is a mixture of protein and lipid membranes. Substances expected to be skin permeable are likely to be those of high water and oil solubility.

There are a number of substances (usually having a molecular weight below 1000) which—if they penetrate the epidermis—can combine with tissue proteins below to

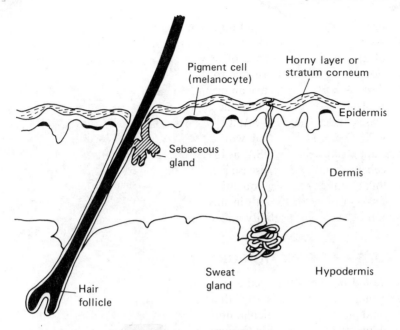

Fig. 1. Basic structure of the skin.

lead to the development of allergic skin disease. Substances which can do this are known as 'sensitizers' and this property is confined to certain substances. The skin contains complex proteins, manufactured by the body, known as immunoglobulins, which can remember substances of this kind and, when encountered at subsequent exposures, can trigger reactions within the skin leading to inflammation and skin disease. Such sensitizers will be described in more detail later in this chapter.

3. OCCUPATIONAL SKIN DISEASES

Confusion can sometimes arise over the use and meaning of the terms, 'eczema' and 'dermatitis'. These are both used to describe the same condition, that of inflammation of the skin, characterized by redness, swelling, blistering, cracking, itching and thickening of the skin. The term 'dermatitis' tends to be preferred in the occupational setting where the cause of the skin problem is often due to contact with a substance in the working environment. Indeed, the major cause of dermatitis within the chemical industry is of this nature and is referred to as 'contact dermatititis'. The term 'eczema' is often used when the inflammation becomes chronic or when the case is thought to be constitutional or familial.

A vast number of substances have been identified as having the ability to produce contact dermatitis if exposure of the chemical to the skin is permitted. Contact dermatitis can result from contact with substances which either have a direct irritant effect on the skin or have the potential to cause an allergic reaction within the skin.

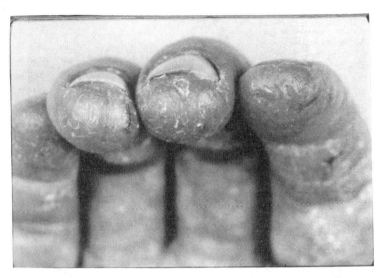

Fig. 2. Irritant hand dermatitis.

The first group causes the condition of 'irritant contact dermatitis' and such chemicals have the ability to damage cells if applied to the skin in sufficient concentration for a sufficient time. The second group causes the condition, known as 'allergic contact dermatitis', and substances able to induce this allergic skin condition are known as 'sensitizers'.

3.1 Contact dermatitis

Any employee having a skin problem and who also has the potential for being exposed to chemicals at work, should be regarded as possibly having contact dermatitis, especially when the hands are affected. There are, however, many other non-occupational skin conditions which may need distinguishing from occupational contact dermatitis, so it should be remembered that contact dermatitis can also result from contact with substances used outside the work environment, in the home or as part of a hobby, etc. (see Fig. 2). Referral to an occupational health physician, the employee's general practitioner and, in some cases, a consultant dermatologist may be required to confirm the diagnosis but there are certain features which can suggest an occupational cause (see Fig. 3).

As mentioned already, the distribution of the skin complaint/rash is important as contact dermatitis is usually confined to the hands and facial areas, at least initially. Enquiry as to when it started, where and what work was being done, may often give clues, particularly if there is a clear relationship between improvement on days off and relapses on return to work. This enquiry should, however, include all potential exposures both at work and about the house or during other pastimes.

Substances which can cause irritant contact dermatitis can vary in their ability to do so, some being mild irritants and others strong. It is the relationship between irritant potential, applied concentration and length of application to the skin which

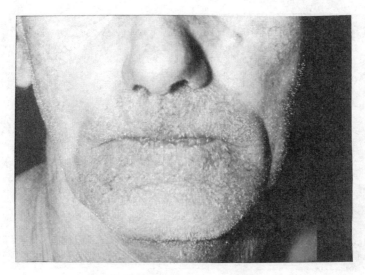

Fig. 3. Chronic irritant contact dermatitis.

is important in the development of the dermatitis. Substances which cause allergic contact dermatitis, i.e. sensitizers, may only require a small amount of the chemical to penetrate the dermo-epidermal junction to induce allergy. However, once the allergic condition has been induced, very small amounts of contact with the same substance may be all that is required to reactivate the dermatitis. Dermatitis may later occur in any area of the skin, not being confined to the area where sensitization first occurred, producing a generalized skin reaction.

It is not possible to say when sensitization will occur in an individual repeatedly exposed to a substance with such a potential. It could be triggered following the first exposure or it could follow, even after repeated exposure with no apparent problem over many years. However, the higher the concentration of exposure, the more likely it is that sensitization will occur. The longer a chemical is in contact with the skin, the more likely it is that a problem will arise (see Fig. 4).

Examples of agents causing irritant contact dermatitis

1. Organic solvents.
2. Soaps and detergents.
3. Cement.
4. Acids and alkalis, such as ammonia and caustic soda.
5. Certain cutting oils and coolants.

Examples of agents causing allergic contact dermatitis

1. Metals, such as nickel, chromium or salts of chromium.
2. Epoxy resins and methacrylate adhesives.
3. Thiuram group chemicals, used in rubber processing.

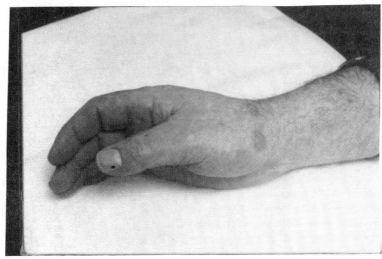

Fig. 4. Allergic contact dermatitis.

4. Sterilizing agents, such as formaldehyde and glutaraldehyde.
5. Certain pharmaceuticals or intermediates, particularly penicillins and cephalosporins.

Some chemicals may have the properties of being both irritant and sensitizing. Examples of these would be cutting oils and certain coolants which can not only irritate but lead to allergic contact dermatitis.

In addition to contact dermatitis, there are a number of other skin problems which can follow exposure to chemicals within the working environment and some of these are worthy of mention.

3.2 Contact urticaria
There are a few substances which, when allowed to come into contact with the skin, produce an almost immediate reaction with the development of redness and blister formation, as typically seen in nettle rash. This allergic hypersensitivity reaction is known as 'contact urticaria'. It tends to be highly irritating and itchy, causes severe discomfort and will recur on repeated contact, often with increasing severity. Typical reaction of this nature can follow exposure to plants such as primula, giant hogweed, etc.

3.3 Vitiligo
Certain substances can destroy the pigment cells within the skin (the melanocytes). This results in pale patches occurring in the skin which become more noticeable when the skin is tanned. This effect can continue and progressively worsen, even if exposure to the offending substance is stopped. Examples of chemicals known to have such potential effects are alkyl catechols, alkyl phenols and quinones.

3.4 Ulceration
Ulceration of the skin can often be found in those who expose their unprotected skin to cement and in those having chronic exposure to hexavalent chromium in, for example, plating shops.

3.5 Acne
There are two specific forms of acne (inflammation of the hair or sebaceous follicles) which have been identified as being due to exposure to substances in the workplace. The first 'oilacne', is the result of mineral oils irritating the hair follicles and 'chloracne', due to inflammation of the sebaceous follicles after exposure to chlorinated aromatic hydrocarbons, such as chlorodibenzodioxins.

3.6 Skin cancer
Prolonged contact with certain substances can predispose to the development of skin cancer in later years. Usually, repeated exposure over long periods of time is required. Examples of substances capable of inducing cancer of the skin are polycyclic aromatic compounds, found in coal tar products, and mineral oils. Indeed, one of the first associations recognized between occupation and cancer was that of cancer of the skin of the scrotum in chimney sweeps exposed to soot.

4. SUSCEPTIBILITY AND PREDISPOSING FACTORS TO THE DEVELOPMENT OF OCCUPATIONAL SKIN DISEASE

Normal skin requires to be adequately hydrated, that is to contain, within the horny layer, at least 10% of water. If the skin is allowed to dehydrate, it will become dry and crack. Work in high humidity environments can result in the skin becoming over-hydrated and soggy. In both cases, the skin will be more susceptible to damage by allowing substances to penetrate more easily.

Thus, it is important, when considering a case of occupational dermatitis, to investigate the whole of the working environment and not simply the substances being handled. With regard to susceptibility to skin irritants, it has been noted from experience that there are certain types of people who may be more at risk—those individuals who have had allergic skin disease (not necessarily occupational) in the past, or have very dry skin or very fair complexion. However, none of these are sufficiently strong reasons to refuse to employ such persons, for excluding them will not prevent skin disease at work if exposure continues to be permitted.

With regard to susceptibility towards sensitizing agents, there is no increased susceptibility in those persons who may have had other forms of allergic disease earlier in life, e.g. hay fever, allergies to cats, dogs or other animals, etc. These individuals are known as atopics and should not be discriminated against with regard to the handling of skin sensitizers.

In summary, therefore, it is not likely to be possible to predict which employees or potential employees are likely to develop occupational dermatitis if exposed. There are two groups of people who, if detected during pre-employment screening, can be considered unsuitable for working with chemicals capable of inducing dermatitis—

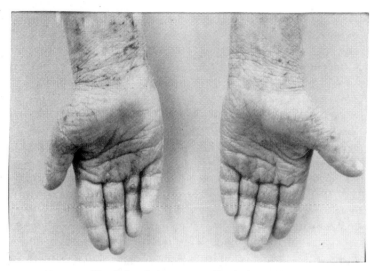

Fig. 5. Atopic (non-occupational) dermatitis.

those currently suffering from allergic skin disease (not occupational) and those with a previous history of a contact dermatitis (whether occupationally induced or not) (see Fig. 5).

The clear message, however, must be that it is by stopping exposure in the workplace that dermatitis will be prevented and not by trying to pre-select a population in the mistaken belief that they will somehow be immune to the chemicals to which they are exposed.

5. PREVENTING OCCUPATIONAL SKIN DISEASES

The aim that we must all share, whether we be occupational health specialists, managers, safety representatives or employees, is to try to eliminate occupationally-induced ill-health and, more specifically with regard to this subject, occupational dermatitis. There are certain issues that, if considered together in a logical sequence, will assist greatly in this objective. This sequence can be summarized as follows:

(a) *Identify* substances which may either irritate the skin or induce sensitization. It is important to consider the pH, solubility (in water and fat) and particle size of substances reviewed.
(b) *Assess* how these substances are likely to be handled in the work environment to enable accurate assessment of the risk of developing dermatitis.
(c) *Monitor* the workforce (health surveillance) and the environment at regular intervals to ensure that problems related to the skin are identified at the earliest opportunity.
(d) *Investigate* any potential skin problem to confirm or refute any association with occupation.
(e) *Notify* cases of identified, occupationally-induced dermatitis to employees, management and appropriate authorities, where necessary.

For readers who like to remember such sequences by using the initial letter of each stage to make a word or words, it is worth pointing out that the above sequence spells out 'I am in'.

It should be the duty of all managers, occupational health specialist, safety officers and others to be available to listen to employees who believe they may have developed an occupational health problem and not dismiss problems without full investigation. Thus, if such an individual tries to contact you, 'I am in' should be the response. There is value in working through these five stages and expanding on how each can help in reducing the risks of occupational dermatitis developing.

5.1 Identifying problem substances

5.1.1 Irritants
The irritant potential of a substance may already be known because of its widespread use throughout industry, from cases of dermatitis which may have occurred or from literature from suppliers. However, new substances will need to undergo certain tests, using animals or, in some cases, human volunteers, to determine the irritant potential. For a substance to be regarded as a skin irritant under the Classification, Packaging and Labelling (CPL) of Dangerous Substances Regulations 1984, it must result in significant irritation to the skin when applied to healthy intact animal skin for up to 4 hours. The inflammation caused should persist for at least 24 hours. It is usual to use albino rabbits for this test, which can distinguish substances likely to have severe, moderate or little potential for being irritant to humans. The test used is known as the 'Modified Draize Test'. Mild irritants can usually be identified by 21-day continuous application tests on animals or humans.

5.1.2 Sensitizers
A test known as the 'Guinea-Pig Maximization Test' is used to distinguish new substances which may have the potential to cause allergic contact dermatitis. However, as described above for irritants, information may already be available on existing chemical substances either from experience or suppliers' information, etc.

This guinea pig test requires a certain concentration of the new substance to be applied to the skin of a specific number of guinea-pigs. (The concentration used is usually the highest tolerated concentration and usually at least 20 guinea-pigs will be tested with an additional 10 control animals.) The initial, high concentration, administered by intra-dermal injection, induces allergic memory in the skin. On subsequent topical application at a lower concentration, inflammation of the skin will result. This test will usually detect most substances which will sensitize humans.

5.1.3 Chloracne/vitiligo/skin cancer
Although not as reliable or readily available, there are tests which can be performed for these potential side-effects of exposure to substances. These are not described further in this chapter, but are as follows.

For chloracne — rabbit ear test or AHH (aryl hydrocarbon hydroxylase) induction test.
For vitiligo — black guinea-pig test.
For skin cancer — rat or mouse repeated application tests.

5.2 Assessment of the working environment

The Control of Substances Hazardous to Health (COSHH) Regulations 1988 can be used as the ideal basis to allow a risk assessment of substances handled in the workplace to be carried out. These regulations are not discussed in detail at this point but it should be noted that the concepts of COSHH are ideally suited towards helping eliminate occupational skin disease (see also Chapter 1).

The risk assessment should take into account the available toxicological data, as previously described, which should indicate potentially hazardous substances. It should then address how these substances will be handled, particularly looking at how exposure (skin contact, in this context) may occur. The following questions should be considered:

(a) Could the substance in question be eliminated or, if not, substituted for something less hazardous to the skin?
(b) Would it be possible totally to enclose the operation so that exposure to the skin could be avoided?
(c) If not, would it be possible to design local exhaust ventilation which would remove hazardous substances from points of potential exposure and thus reduce the likelihood of contact with the skin?
(d) If none of the above is possible, has adequate personal protective equipment been provided?

In addition, as also required by the COSHH Regulations, it is vitally important to make regular checks on exhaust ventilation systems and personal protective equipment to ensure their degree of function and integrity. It is also vitally important to train employees involved in work with hazardous chemicals in good housekeeping and safe working practices and to ensure that management supervision is both strong and conscientious with regard to all aspects of safe working. It is perhaps above all else the risk assessment of the workplace which is likely to have a major impact on reducing occupational ill-health.

5.3. Monitoring

Both the workforce and the workplace should be monitored to identify cases of occupational dermatitis and to ascertain the levels of substances in the working environment. Monitoring the health of the workforce is known as 'occupational health surveillance' and monitoring the work environment is known as 'occupational hygiene monitoring'. (More information on the latter is given in Chapter 11 of this book.) However, it is worth re-emphasizing some of the points which have particular relevance in preventing occupational dermatitis.

5.3.1 Health surveillance

There are a number of substances which, if used in a manufacturing process, require health surveillance under the supervision of an Employment Medical Adviser (EMA) or Appointed Doctor (AD). Examples of these are as follows:

— substances known to cause severe dermatitis,
— any process using vinyl chloride,
— the manufacture of patent fuels from pitch and
— chrome solutions in chrome plating, dyeing and tanning.

In addition, the COSHH Approved Codes of Practice on the Control of Carcinogenic Substances and Fumigation Operations identify carcinogens and fumigants as substances which require statutory health surveillance.

It is the COSHH assessment which should be used to identify potential exposures, likely ill-effects and, therefore, the need to perform health surveillance for substances not covered by statutory surveillance. This is particularly relevant for new chemical entities, for which there is little experience of handling.

The aim of health surveillance with regard to occupational dermatitis is to identify cases of work-induced skin problems as early as possible to prevent the development of a chronic condition which may persist even when exposure to the offending agent ceases. It also identifies the need to improve containment of substances at source and allows early treatment for the individual affected, with a reduction in illness and rehabilitation time and, possibly, less likelihood for the need to consider redeployment. Health surveillance for occupational skin disorders need not necessarily involve regular examination by a medically qualified person, although it will be usual to require the advice of an occupational health physician or medical practitioner with relevant experience in the first instance. Such an individual will be able to advise on the frequency and degree of surveillance required.

Methods of health surveillance available
(a) *Self-examination*

The first person to notice any change in the skin is likely to be the individual concerned. If employees have the confidence that their management will investigate problems without threat of termination of employment, they will usually be willing to report problems early, especially if there is a member of the occupational health department in whom they have confidence and to whom they can report. If assessment of a manufacturing operation suggests that there is a likelihood of skin problems developing in the operators, they should be briefed on this and instructed to report any skin changes to their appropriate line manager and occupational health department.

(b) *Examination by a responsible person*

One does not have to be a qualified doctor or nurse to be able to recognize the changes in the skin which may result from exposure to chemicals in the workplace. Employers can appoint suitably motivated personnel to be 'Responsible Persons' who, following training and instruction from a doctor or nurse, can be used to monitor the workforce for skin problems. A good example would be to look for ulceration of the skin due to exposure to chromium solutions. Such a responsible person could simply routinely enquire about problems or, in some instances, physically inspect individuals' skin. This would depend on the perceived risk and specific substances to which they have been exposed. Responsible persons should have the back-up of either a trained occupational health nurse or physician.

(c) *Examination by a nurse or doctor*
For non-statutory surveillance, this is likely to be available only in larger companies having established occupational health nurses or occupational health centres.

5.3.2 *Monitoring the workplace*
Before discussing how the working environment should be monitored for substances which may cause occupational skin problems, it is first worth considering the most common ways in which the skin of workers may become contaminated by substances within the working environment. Within the chemical industry, exposure to potentially harmful substances can occur either directly or indirectly. If dusts, fumes, vapours, etc. are not controlled at source, they will contaminate the work environment and come into contact with areas of exposed skin, which usually means the hands and face. Contamination of tools, work surfaces, etc. will lead to direct contamination of the hands. Indirect contamination may occur from contaminated hands/gloves touching the face. Face touching is an almost universal phenomenon and is usually done unconsciously. The face can thus be very easily contaminated from the hands/gloves unless the individual makes a deliberate effort to prevent this. Contaminated surfaces may not always be obvious and so it is of vital importance to control exposure at source.

It therefore becomes apparent that the two major sites for occupational skin problems are the hands and face. With regard to the face, the tissues around the eyes (periorbital) are particularly thin and vulnerable both to irritant and sensitizing agents and also to the defatting properties of solvents. Periorbital redness, dryness and flaking of the skin is the most common type of facial skin problem. With regard to the hands, various degrees of reddening, blistering, cracking, drying and thickening of the skin are the usual presenting features of skin exposure to chemical agents in the workplace, whether they be irritants or sensitizers. Tests are available to distinguish irritants from sensitizers and these are discussed below.

Thus with regard to monitoring the working environment, consideration should be given to monitoring airborne dusts, fumes and vapours. In addition, surface swab testing can give an indication of the extent of contamination of surrounding plant if containment at source has not been total.

By using combinations of static and personal dust/vapour/fume monitoring results, a detailed picture can be built up of a workforce's exposure and comparison with occupational exposure limits can be made. It should be noted, particularly with regard to sensitizing agents, that it is the acute, short-term exposure which is often of more relevance to the subsequent development of ill-health effects.

Swabbing surfaces to look for contamination is a valuable tool. Swabbing over a set surface of 10 dm^2, although not directly related to the amount absorbed through the skin, will give an indication of the degree of contamination of the surface and can be used serially to see if control measures are actually working.

Environment monitoring is carried out as an aid to assessing the control of substances hazardous to the skin and other parts of the body. In doing so, it can identify processes requiring improvement to reduce the likelihood of occupationally induced skin problems. Under the COSHH Regulations, all reasonably practicable efforts to prevent exposure should be carried out before relying on the use of personal protective equipment. However, if total containment is not possible, the need to use and select appropriate equipment will be apparent. However, equally important is the need to train the workforce in its use, in good manufacturing and housekeeping practices and to ensure good management supervision. Reference should be made to relevant safety and occupational hygiene literature for further advice on protective equipment.

5.4 Investigating possible occupational skin problems

The main factor which is likely to lead to the successful detection of an occupational skin problem is having a high index of suspicion, that is, being constantly on the look-out for skin problems and asking 'Could this be occupational?' However, having developed the desire to look for cases, it is necessary to understand how one should then proceed to investigate.

Often a single case of a skin problem (the index case) may herald the arrival of a number of similar cases (clusters). These, however, may need to be actively sought. Thus, if an index case should indicate a possible occupational problem, all other similar employees should be interviewed and, if necessary, examined to estimate the extent of the problem. This is known as carrying out a dermatological survey.

If there is any doubt as to whether the problem is occupationally induced, advice should normally be sought from an occupational health physician or dermatologist. In difficult cases, there would be a need, in such surveys, to identify and match exposed groups of individuals with control groups and advice will be required on epidemiology and statistical analyses of the findings of any study. All such surveys should be done with the full understanding and knowledge of the workforce and results explained to them in a manner which they can easily understand.

When cases have been referred to medical practitioners with appropriate specialist skills, there are a number of further investigations which can be carried out, following history taking and examination that will help the doctor to reach the correct diagnosis. These are as follows:

(a) skin prick testing,
(b) patch testing,
(c) blood tests,
(d) skin biopsy.

They are specialist investigations and are, therefore, only briefly described here.

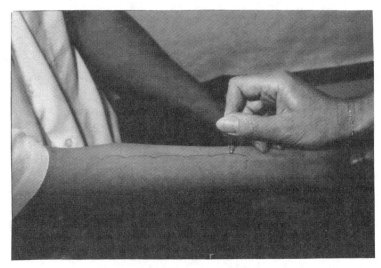

Fig. 6. The skin prick test.

5.4.1 Skin prick testing

In this test, a pre-determined concentration of the agent suspected of causing the occupational skin problem is injected, in solution, superficially into the tissues of the skin of the forearm. If the test is positive, i.e. the subject is allergic to the substance being injected, a wheal will appear, surrounded by an area of redness (as seen in cases

Table 1. Notifiable skin diseases

Skin disease	Work activity	Common description of disease
Chrome ulceration of (a) the nose or throat (b) the skin of the hands or forearms	Work involving exposure to chromic acid or any other chromium compound	Chrome ulcer
Folliculitis	Work involving exposure to mineral oil, tar, pitch or arsenic	Folliculitis
Acne	" "	Acne
Skin cancer	" "	Skin cancer
Inflammation, ulceration or malignant disease of the skin	Work with ionizing radiation	Radiation skin injury

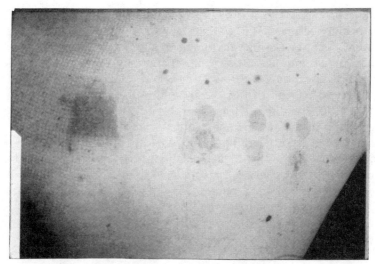

Fig. 7. Patch testing for allergic contact dermatitis.

of nettle rash). This test is used in cases of suspected contact urticaria and contact dermatitis due to protein complexes. It is an indication of allergic disease due to the presence of certain antibodies to the agent in question. In this case, they are known as immunoglobulin E (or IgE for short). The test takes only 15 minutes to complete (see Fig. 6).

5.4.2 Patch testing
In this test, a pre-determined concentration of the agent suspected of causing the occupational skin problem is applied to the skin, either in solution or mixed with white soft paraffin/vaseline. The site chosen is usually the back, and the agent in question is covered with a patch for usually 48 hours. If the test is positive, this results in a reddening of the skin below the patch. This test is used to distinguish between allergic contact dermatitis, in which case the test will be positive, and irritant dermatitis, in which case the test will not. It will also distinguish allergic contact dermatitis from skin disease that is not occupationally induced (see Fig. 7).

5.4.3 Blood tests
It was mentioned above that skin prick tests, are an indication of IgE present in the skin. This Immunoglobulin antibody can also be measured in the blood itself. Again, this is used in cases of contact urticaria but is a more invasive technique. Other tests are available which are even more sophisticated, e.g. the RAST (Radioallergosorbent) test. Here, specific antibodies to agents causing allergic skin disease can be marked, following a blood sample being taken, with a radioisotope and thus quantified. This test is used in the detection of protein contact dermatitis.

5.4.4 Skin biopsy
Histological examination of the lesion can be helpful in identifying the cause of lesion in some cases.

5.5 Notification

Having established that a skin condition may be caused by exposure to a chemical agent from within the working environment, it will be necessary to consider whether the condition falls into the category, of a 'Reportable Disease'. Under the Reporting of Injuries, Diseases and Dangerous Occurrences Regulations (RIDDOR) 1985, there is a requirement to report to the local office of the appropriate Health and Safety Inspectorate the skin diseases listed in Table 1. If more than one person is affected by a reportable disease, a separate record should be made for each person.

With improved reporting of such skin problems, a much greater understanding of the incidence and prevalence of such conditions within, industry could be obtained. However, the list does not cover contact dermatitis arising from all substances encountered at work and, therefore, fails to collect sufficient information on this significant group of occupational health problems.

6. MANAGEMENT OF CASES OF OCCUPATIONAL SKIN DISEASE

Prevention is always better than cure but, having failed to prevent occupational skin diseases, it is important to understand the principles of management of individual cases. What should be done when a case is diagnosed as having an occupational cause and how are the individual's symptoms likely to develop? There are a number of factors that can affect the outcome of the skin problem and these are discussed below.

Firstly, it is usual that the longer someone has suffered from an occupational skin problem, e.g. contact dermatitis, the longer it takes for the problem to resolve. With regard to irritant dermatitis, i.e. contact dermatitis, if exposure is stopped or significantly reduced, the problem, in most instances, resolves. Sometimes, however, for some unknown reason, the dermatitis can persist for a long time, even if exposure ceases. Thus, it becomes obvious that, in cases due to irritant dermatitis, providing the cause is identified and controlled, employees should be able to return to their previous occupation. However, the odd individual with persistent problems should seek alternative work.

Allergic contact dermatitis arising from certain industrial chemicals usually requires redeployment away from further exposure. Normally, the amount of chemical required to perpetuate the problem is much smaller than the amount required to cause the initial allergy. Controlling environmental levels of the chemical agent in question to levels below those which may initiate the allergic process may not be sufficient to stop the problems associated with the developed case.

If individual cases of dermatitis show a reluctance to heal, despite adequate control or redeployment, consideration should be given to any possible aggravating factors. The common aggravating factors are as follows:

(a) Infection—if the dermatitis becomes infected, it may be slow to resolve unless antibiotics are used.
(b) Allergy to treatments—it has recently been recognized that allergy can even develop to steroid creams and ointments used to treat dermatitis. If this additional allergy develops, treatment may aggravate the condition and delay healing.

(c) Allergy can develop to the rubber in rubber gloves, used to prevent exposure to chemicals and can be another reason for failure of a dermatitis to improve on stopping exposure.

The following are some practical steps which should be taken when the cause of the occupational skin problem has been identified.

(a) Informing the employee of the nature of the problem
The employee should be informed as to the precise cause of his problem and to avoid further exposure. This may require some retraining and use of additional protective equipment, etc., if it is not possible to eliminate exposure by control at source.

(b) General advice on aggravating factors
This should be given so that the employee can be made aware of the factors—from both the work and the home environments—which can delay the healing of the symptoms. The care of the skin should be emphasized along with the benefits of using barrier creams, skin cleansers and conditioners. The need for careful drying of the skin after washing is also important.

(c) Continued work versus redeployment
The individual concerned should be informed of the relative merits or risks of continuing in the same employment or in seeking redeployment. If redeployment is considered necessary, every effort must be made to redeploy within the same workplace. If this policy is not followed, there is a significant risk that employees will not report their skin problems early for fear of dismissal, with resultant loss of livelihood and socio-economic status. Individual cases will, in most instances, require advice on this issue from a dermatologist or experienced occupational health physician.

7. CONCLUSIONS

Occupational skin problems are avoidable, not by pre-selecting employees considered to be less susceptible but by controlling exposure to chemicals at source. As soon as a condition has been suspected, by appropriate health surveillance, it should be thoroughly investigated to identify the precise causative agent or agents. Other cases should be sought and those affected treated sympathetically, with full explanation of the problem, and every effort made to maintain employment and socio-economic status.

14

Chronic health effects and chemical control

Dr P. A. Martin

SUMMARY

This chapter is aimed at providing the reader with an appreciation of the many factors which play a role in the toxicological outcome following chemical exposure.

The introductory sections attempt to explain the basis of the science of toxicology with its central hypothesis of the dose–response relationship, the competing effects of metabolism (activation *vs.* detoxification) and also set out the concepts of hazard and risk.

The chapter goes on to develop the theme of the need for chemical control programmes, at a European level as well as at national level, following the rapid expansion in industrialization after the Second World War, and the realization of the wider environmental perspective. This caused the EC to institute environmental programmes and to begin to address the need for a hazard/risk assessment for *existing* chemicals as well as new chemicals.

The public perception of chronic adverse health effects following exposure to chemicals, exemplified by cases of occupational ill-health, is mainly focused on diseases such as cancer, nervous system impairment and birth defects. This later section concentrates on carcinogens, mutagens and teratogens since toxicological testing protocols are widely harmonized and accepted in these areas.

Having derived relevant toxicological test data one then has to utilize this to make judgements about the amount of a chemical to which humans may be exposed continuously, during their working life, and which will not induce adverse health effects. This process is called risk assessment. Two approaches are described.

The final section of the chapter aims to convince the regulators and governments of the need for more involvement and openness in the debates which take place prior to the setting of permissible exposure limits or discharge levels and to reassure the public of the integrity of scientists who are called upon to make these judgements. They are only human; the data from which the judgements are derived are not infallible. Life itself involves taking risks; there are no guarantees of absolute safety.

1. INTRODUCTION

The classical definition of toxicology is 'the study of the adverse effects of chemicals on living systems'. There is nothing new about the fact that chemical substances, be they derived from natural products or synthesized, may, under certain conditions, cause adverse health effects.

Paracelsus (1493–1541) coined the now well-used phrase, 'All things are poisons, for there is nothing without poisonous qualities'. This revolutionary view of Paracelsus' went on to direct certain integral parts of the present-day study of toxicology. That is:

(a) experimentation is a pre-requisite in the investigation of the responses to chemical action,
(b) one should clearly distinguish the therapeutic from the toxic properties of substances,
(c) these properties may be separated by the dose required to elicit the response
(d) a certain degree of specificity of action may be delineated at therapeutic or toxic doses.

In many respects, Paracelsus articulated, in these statements, the central theme of the science of toxicology and, today, they form the basis of the dose–response relationship.

From its earliest beginnings, toxicology, as the study of poisons, developed alongside, and in parallel with, pharmacology. In recent decades, however, the definition of toxicology has been broadened so that the two disciplines have now been irrevocably separated. Nowadays, toxicology is viewed as the study of adverse effects of chemicals on living organisms in order to predict the chemical hazards to humans. Further expansion of the definition to include 'chemicals and/or other agents' together with the addition of 'man or his environment' followed to represent our present understanding of the science of toxicology.

Toxicology is utilized as an ancilliary to public and/or community health and preventive medicine and is expected to enable a prediction to be made of the likely effects of a chemical in a complex biological system. By analysis of the dose which causes these effects and the use of a safety factor/margin, it is held that exposures to the chemical in a variety of applications may continue. These safety factors/margins are judgements based upon findings in experimental systems of 'no observed effect levels' (NOELs) and are widely used in safety regulations aimed at the control of chemicals. However, the toxicity of a chemical is not just an innate constitutive property of the molecule; toxicity is highly dependent upon the conditions under which living organisms are exposed to the chemical (see Table 1).

Table 1. Some factors affecting the biological response to chemicals

Amount of the chemical
Physical and chemical properties of the chemical
Route of exposure
Absorption of the chemical
Bio-transformation of the chemical—distribution, metabolism
                                              accumulation and elimination.
Species
Age and sex of the exposed animal
Health and dietary status

2. HAZARD AND RISK

An important factor when considering the toxicity of a chemical in relation to any practical expression of its ability to cause harm is the distinction between hazard and risk.

Hazard is best defined as the property of a substance to cause toxic effects if a living organism (human or experimental animal) is exposed to it. It is a manifestation of a biological property of the material in relation to some practical event and, as such, is open to experimental investigation.

Risk is the predicted or actual consequence of exposure in a defined population to a given dose or concentration of the material. Risk represents the frequency and intensity of the toxic actions in a given group of people, animals etc., under a defined set of circumstances.

These definitions are consistent with those recommended by the Organization for Economic Co-operation and Development (OECD). They permit the toxicologist to define the limits of experimental studies and point to areas where data are required to make a 'real world' prediction of the consequences of exposure rather than those under controlled laboratory conditions (see Fig. 1).

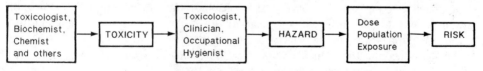

Fig. 1. Diagrammatic representation of an assessment of 'Chemical Safety'.

Relatively few chemicals are toxic *per se* and these generally manifest acute toxicity, that is, they react directly with tissues or cause tissue reactions, such as narcosis (e.g. alcohols) or acetylcholinesterase-induced paralysis (e.g. organophosphates). Many chemicals are metabolized into reactive intermediates (see Table 2) which then may react directly with tissue proteins, enzymes or DNA etc., to cause acute lethal injuries or chronic toxicity, such as reproductive effects and tumorigenicity.

Table 2. Some examples of activation of chemicals to toxic entities

Chemical class	Example	Metabolic reaction	Toxic metabolites	Toxic effects
Haloalkanes	CCl_4	Reductive dehalogenation by microsomal cytochromes	Trichloromethyl radical and active oxygen	Hepatotoxicity
Aromatic hydrocarbons	C_6H_6	Oxygenation by microsomal cytochromes	Benzoquinones and active oxygen	Aplastic anaemia and cancer
Thio-phosphonate compounds	Parathion	Sulphur-replacement by microsomal cytochromes	Paraoxon	Neurotoxicity

3. CHRONIC TOXICITY

Not all exposures to chemicals result in toxicity and adverse health effects. At least four inter-related steps spanning the presentation of the chemical at a route of entry into the body, through to the clinical manifestation of an effect on the tissue/organ of the body, are involved in the eventual outcome—plus or minus toxicity.

This is best illustrated by reference to Fig. 2 (adapted from [1]).

Adverse health effects which result from exposure to a chemical, and which are termed chronic toxicity, may be the result of either:

(i) prolonged or repeated administration of the chemical prior to manifestation of an effect, or
(ii) an adverse effect which is long-lasting but which may be caused by one or very few exposures to the chemical.

(i) The following examples illustrate some ways in which a chemical, by repeated or continuous exposure, at low concentrations, may exert its biological effects on the target tissue:

(a) *A build-up of a metabolite.* For example, cadmium—exposure to low levels of cadmium over a number of years may lead to kidney damage. This may be diagnosed through the appearance in the urine of a specific, low molecular weight protein. Tissue repair may occur in the kidney cortex cells and thence recovery if exposure ceases; otherwise, the disease tends to be progressive.

(b) *By causing cell necrosis.* Carbon tetrachloride can kill certain liver cells called hepatocytes. Following exposure to low concentrations of carbon tetrachloride, a relatively small number of hepatocytes may be killed. The remaining cells will divide in an attempt to replace the damaged cells—a process referred to as hypertrophy. However, another consequence of this process involves other structural elements of the liver. These cells are activated or switched on and start to lay down connective tissue or collagen. This, in turn, reduces the functional capacity of the liver and is referred to as cirrhosis of the liver.

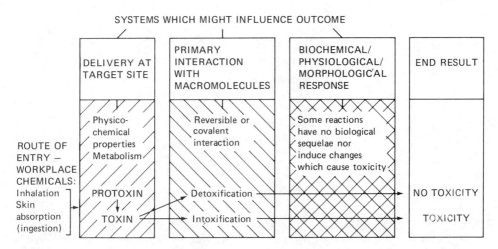

Fig. 2. Some factors which influence toxicological outcome following exposure to a chemical.

(ii) One of the best researched examples of this type of chronic effect is referred to as delayed neuropathy and may be caused, for example, by exposure to certain organophosphorus compounds. Delayed neuropathy is characterized by an unsteady gait and may progress to eventual paralysis. Although exposure to the chemical may be for a very short duration, it can initiate a 'biological cascade', leading to significant adverse health effects at some future date.

The target within the body is the nervous system. Here, enzymes—neurotoxic esterases—are attacked by the organophosphorus compound and chemically altered. This alteration is thought to cause changes to the surrounding membranes to which these enzymes are intimately bound. If these damaging effects occur in peripheral nerves, some repair and return of function may take place. However, no such repair occurs in nervous tissues of the spinal cord.

This example illustrates the hypothesis that certain chronic diseases, including cancer, may be initiated by a chemical interaction which occurs long before there is any clinical manifestation of a disease or adverse health effect. This 'lag period' between exposure and appearance of disease/ill health is often referred to as the latent period. Certain biological systems appear to be 'switched on or off' and the ensuing changes, once initiated, do not appear to be reversible.

The usual aim of a chronic toxicity study is to characterize the profile of the effects of the test substance in an appropriate mammalian species following prolonged and repeated exposure. The duration of chronic toxicity studies for effects other than cancer (neoplasia) is subject to continuing debate. However, it is reasonable to assume that a chronic study is one carried out over a period in excess of 4 months and may be as long as 2 years, depending upon the experimental model. Under these conditions, effects such as carcinogenicity, which usually require a long latent period, or cumulative adverse effects, should be manifest.

Data obtained from these studies should allow for the identification of the majority of chronic effects and also the demonstration of any dose–response relationship. In addition, by appropriate experimental design and study conduct, these investigations should allow for the detection of aspects of general toxicity, including neurological, physiological, biochemical and haematological effects, as well as exposure-related morphological (pathology) effects.

The results of chronic toxicity studies should allow permissible levels of exposure (highest permissible concentrations) to the substance in the air of the workplace to be set [2]. In other branches of experimental toxicology, the assessment of toxic effects evoked by chronic exposure should permit the setting of acceptable daily intakes (ADIs) or other safe limits for the chemical in food and water.

4. THE CONTROL OF INDUSTRIAL CHEMICALS

4.1 Introduction
Chemicals are essential for the maintenance of today's expectations of 'living standards' and thus they cannot expect to escape the considerations of politicians and regulators. Following the Second World War, there was a continuous expansion of international trade in chemicals, especially among industrialized countries. This expansion slowed in the 1980s and 90s and manufacturers of industrial chemicals have sought to maintain their competitiveness by searching for markets in less developed countries.

Safety evaluation of chemicals is fundamental to the development of environmental policies. During the 1980s, the need for international co-operation to achieve any sustainable environmental protection, in its widest context, was increasingly realized and this led, for example, to the signing of the Montreal Protocol to curtail the production of chlorofluorocarbons (CFCs) and hence reduce the threat of worldwide ozone depletion.

European Community environmental policy began in earnest in 1973 with the adoption of the first 5-year Environmental Action Programme. However, certain elements of chemical control had been in operation for some time prior to this. In the international community, the Organization for Economic Co-operation and Development (OECD), which presently comprises countries from N. America, Europe, Asia and Oceania, formed an Environment Committee. By 1979, a 'Management Committee for the Special Programme on the Control of Chemicals' had been established which, together with the principles of the 'action programmes' of the EC, have led to tremendous progress in the control of chemicals in the intervening years. Within the EC, this has included the adoption of over 100 legal instruments encompassing a number of key directives.

4.2. The 'Dangerous Substances' Directive
The Sixth Amendment to Council Directive 67/548/EEC on the Approximation of the Laws, Regulations and Administrative Provisions relating to the Classification, Packaging and Labelling of Dangerous Substances (CPL) was developed in close association with the OECD in order to achieve international agreement on the elements of a system of *preventive control*, primarily for 'new chemicals'. This amendment, like most EC environmental legislation, has the following fundamental aims:

(i) to create harmonized conditions of marketing, thereby ensuring the free movement of products in trade within the EC, and
(ii) to protect public health and the environment against the potential hazards of chemicals.

The Directive essentially regulates *industrial* chemicals in the following areas:

(a) **Dangerous Chemicals**—listed in Annex 1 of the Directive,
(b) **Existing Chemicals**—present on the EC market prior to September 1981 and listed in EINECS (the European Inventory of Existing Commercial Chemical Substances) and
(c) **New Chemical Substances**—not in the two lists above and which need to undergo a 'notification procedure' prior to being placed on the market.

Over the past 2 years or so, there has been considerable discussion within the European Commission, based upon experience gained by competent authorities in member states during the operation of the Sixth Amendment. This has led to the proposal for a Seventh Amendment [3], through which the Commission seeks to introduce improvements to current legislation on control of chemicals and the reduction of adverse health and environmental effects from the manufacture and supply of chemicals. These will include, amongst others, a classification category and labelling criteria for 'Dangerous to the Environment', a modified and harmonized system for notification of new substances, to replace the 'Limited Announcement' schemes for small volume chemicals, and a section to reinforce the principles of 'reduction, refinement and replacement', in an attempt to curtail the need for unnecessary animal experimentation.

Many other facets relating to the control of chemicals are also reviewed in this latest amendment. Chemical manufacturers hope that this new legislation may help to resolve the issues relating to risk assessment and risk management which have been inadequately addressed by previous legislation. The latter sought to control chemicals solely on the basis of the hazards they presented by virtue of their *intrinsic* properties. The Seventh Amendment introduces a requirement for risk assessment to be produced prior to the introduction of new substances on to the market. These risk assessments will have to be carried out in line with general principles laid down in an EC Directive on Risk Assessment for New Chemicals, due for adoption in April 1993.

4.2.1 *'New' chemicals*
Under UK law, the Notification of New Substances Regulations 1982 came into force on 26.11.82 and embodied the notification provisions of European Directive 67/548/EEC and its various amendments relating to:

(a) its provisions concerning the classification, packaging and labelling of dangerous substances and
(b) the introduction into the EC of a common notification scheme for new substances.

These Regulations were amended in 1986 and 1991 and revised in 1988 [4] in order that UK legislation was harmonized with the requirements of various related Community directives.

A new substance is defined as:

> 'any substance except a substance which had been supplied (either alone or as a component of a preparation) in a member state at any time in the period 1.1.71 to 18.9.81 (that is not on EINECS)'.

A substance may be supplied by a manufacturer/importer *alone*:

> 'either as a chemical element or compound, in its natural state or as produced by industry, and includes any additives such as stabilisers, inhibitors and impurities that are part of the final marketable substance, or

as a component of a mixture

> where the mixture is the only form in which the new substance is manufactured and the new substance cannot be separated for technical reasons from this mixture'.

The type and extent of the information that is required depends upon the quantity of the new substance produced and includes basic physico-chemical properties of the material, such as structural/empirical formulae, solubility (in both fat and water), vapour pressure and also details relating to flammability and explosive properties. In addition, there is a requirement for a wide range of toxicological and ecotoxicological tests to assess the new chemical's likely biological impact on human health and the environment and also to provide sufficient information for classification and labelling purposes. Some further information on applications, together with environmental fate and dispersion, may be required in order to conduct an adequate risk assessment under the Seventh Amendment.

The tests required under the Notification Scheme are often referred to as the Base Set and may be placed into the following categories for administrative convenience:

1. Investigation of physico-chemical properties.
2. Investigation of acute toxic effects.
3. Investigation of skin and eye irritation and sensitization.
4. Investigation of sub-acute effects.
5. Screening tests for mutagenic and potential carcinogenic activity.
6. Investigation of acute aquatic toxicity and biodegradation potential.

For full details of the tests, the reader is referred to the Approved Code of Practice, the OECD Test Guidelines and the Annex V Test methods [5–7].

Information gained from the categories of tests described above form a *minimum* data source from which an informed judgement may be made on likely human and environmental hazards following exposure to the chemical tested. Additional information on toxicological and ecotoxicological parameters may be required as the manufacture and supply tonnages reach certain pre-determined threshold levels. For example, the competent authority may require certain additional test data (referred to as Level 1 test data) when the quantity of substance placed on the market reaches a level of 10 tonnes per year or a total of 50 tonnes. These additional tests may include reproductive toxicity, additional mutagenicity and ecotoxicology studies. A second threshold level of 1000 tonnes per year or a total of 5000 tonnes also requires that the manufacturer/importer supplies further test data to the regulatory authority.

4.2.2 Existing chemicals

'Old' or 'existing' chemicals have been the subject of environmental regulation for specific media (such as air, water and soil) or the control of industrial emissions and operations for a number of years. In addition, since 1980 there has been, considerable increase in the gathering and interpretation of data relating to existing chemicals on an international scale. These international efforts have been led by the OECD and, more recently, the European Commission which published a draft proposal for a Regulation on the Evaluation and the Control of the Environmental Risks of Existing Substances [8] in December 1989. Both programmes are driven via lists of identified 'priority chemicals' which will form the basis for the legislation to require additional test data. A decision, a risk assessment, based upon an evaluation of these data, will enable any necessary regulatory actions to ensue. Within the European Community, this regulation/control is most likely to take the form of either a total ban or restrictions on the marketing and use of a substance.

The 'priority chemicals' have been selected primarily on a production volume basis (see Fig. 3). However, at this stage, the two programmes differ in that: within the OECD programme there was a panel of experts who examined these chemicals—so-called high production volume (HPV) chemicals—and 'screened out' those of limited further concern at this stage. Within the EC, the legislation requires a dossier of information on *all* HPVs, some 1,500 chemicals (as listed in Annex 1 of the draft Regulation), which places onerous demands on the European chemical industry.

Phase 1:	> 1000 tonnes on Annex 1	12 months from adoption Full dossier*
Phase 2:	> 1000 tonnes on EINECS	18 months from adoption Full dossier*
Phase 3:	10–1000 tonnes on EINECS	4.5 years from adoption Limited dossier*

Fig. 3. Proposed schedule of dossier submission on existing chemicals. *For further details on dossier requirements, see [8].

The second factor considered in the generation of these priority lists was the available toxicological and environmental data on these HPVs identified from Stage 1. In most cases, it was anticipated that for 'priority chemicals' there might as well be some deficits in the currently available knowledge which would preclude any assessment of potential risk. In order that a valid risk assessment might be made, these 'data gaps' would need to be filled.

The Regulation is very extensive in its scope and, by means of a phased approach, it is anticipated that risk assessments for existing chemicals (in line with a proposed EC directive on 'Risk Assessment for Existing Chemicals' produced in quantities down to 10 tonnes will be addressed within 5 years of the Regulation being adopted. However, the dossier of information required will be restricted to substances in the lower production brackets, that is, between 10 and 1000 tonnes.

This concept of risk assessment needs to be integrated into regulatory decisions governing the control of chemicals, since it is only when exposures are taken into account that we will be better able to refine some of the crude decisions which have hitherto been made with respect to the CPL of substances which were based upon the intrinsic properties of the chemicals.

EINECS contains some 100,000 chemicals and thus the extent of this programme is immense. The aims of both the OECD and the European Commission in this area are essentially similar although the 'prioritization' and the data elements required in the dossier submitted are not fully harmonized. Nevertheless, data submitted to one programme will, hopefully, be fully acceptable to the other.

5. TOXICOLOGICAL INVESTIGATIONS FOLLOWING CHRONIC EXPOSURE

5.1 Introduction

The public concern with respect to adverse health effects following exposure to chemicals in the workplace has focused mainly on chemicals which induce irreversible conditions such as cancer. This section will deal primarily with chemicals which are referred to as carcinogens, mutagens and teratogens within the Sixth Amendment (see previous sections). Although other important chronic adverse health effects, such as neurotoxicity and immunotoxicity, are known, at this time there are no fully harmonized and accepted routine test guidelines in these areas.

5.2 Mutagens

In considering the nature and origin of cancer, it is necessary to take account of the structure and functioning of living cells. Cells are composed of complex molecules which are organized in a well-structured manner so as to carry out the many and varied reactions which occur within and between them. One of the most important groups of these complex molecules are the proteins.

Proteins are polymers composed of specific arrangements of 20 different monomeric units—the amino acids—which are linked by peptide bonds. The activity and behaviour of a particular protein depends on the number and the arrangement of the amino acids from which it is constituted.

Despite their importance to living cells, proteins do not have the capacity to reproduce themselves. The information and instructions for this are stored in the cell in the form of sequences of nucleotide bases in a molecule called deoxyribonucleic acid (DNA). DNA contains, in its structure, a template for all proteins in an organism and is responsible for carrying all the information necessary for the growth and functioning of all cell types. DNA is the means by which genetic information is passed from parent to offspring—heredity.

The hereditary information within DNA is organized in units which specify the sequence of a particular protein. These units are called genes. In simple living organisms, such as bacteria, viruses and blue-green algae, the genes are carried by a single molecule which may be circular or linear. In these simple prokaryotic cells, the DNA is contained in a regular structure called the nuclear body which is found in the cell cytoplasm. In more complex organisms, referred to as eukaryotes, such as human beings, the genes are distributed along a number of different linear pieces of DNA called chromosomes. These chromosomes are enclosed within the nuclear membrane except at the time of cell division.

When cells multiply (by dividing in two) the genetic material is duplicated so that, with cell division, this material, and thus the information and instructions, is passed to each of the daughter cells. Each cell should therefore contain a complete set of instructions following cell duplication. In body or somatic cells this process is called mitosis. However, early in the development of an embryo, a groups of cells, the germ cells, are set aside to be the precursors of the future gametes (ova or spermatozoa).

Germ cell development is distinct from that of somatic cells. Each mature gamete only contains half the number of chromosomes of a somatic cell—a haploid set of chromosomes. This process of division is known as meiosis. The fusion of male and female haploid gametes results in the production of a diploid stage which, with subsequent mitotic cell divisions, produces a new individual for the next generation.

Mutations are changes to these instructions and arise through alterations to the DNA at the time of cell division. These changes occur as a result of alterations within a gene (causing alterations in the protein for which the gene codes), or to events involving chromosomal changes, or even losses of pieces of information. Once a mutation has occurred, it is likely that this change will be perpetuated each time the cell divides unless the 'damage' is repaired. Living systems have evolved a battery of repair processes effected via enzymes to 'correct' lesions on strands of DNA.

Experimental investigations involving the use of ionizing radiations and a variety of different chemicals has led to a growth in the understanding of the interactions between changes in the cell DNA and the induction of cancer. Over the past 20–30 years, a variety of tests have been developed which attempt to assess the ability of a test material, usually a chemical, to cause heritable or transmittible generic changes. These latter mutational changes are thought, by most scientific authorities, to be the fundamental first step in the process of cancer induction, or carcinogenesis. However, it should be borne in mind that cancer is a multi-stage disease and that mutation of genetic material is probably only one, albeit crucial, stage in a series of events leading to tumour development. Our current understanding of the mechanisms of tumour development—cancer—indicates that any chemical which induces mutations in mammalian cells should be regarded as being a suspect carcinogen. Tests using bacterial and mammalian cell systems have been developed as 'predictors' of the interaction of chemicals with DNA—these tests are generally referred to as short-term tests.

Broadly speaking, genetic alterations in eukaryotic cells may arise from three categories of mutational events, which occur:

(i) at the gene level,
(ii) at the individual chromosome level or

(iii) at the level of the chromosomal set, that is, a change to the normal chromosomal number.

These three categories form the basis for a 'testing battery/strategy' promoted by the Department of Health for the investigation of the mutagenic properties of a substance [9]. These guidelines suggest that virtually every chemical with mutagenic potential could be detected if subjected to a set of three *in vitro* assays (see Fig. 4). The following tests are routinely recommended:

(i) a bacterial cell assay for gene mutation—this would most usually be the Ames test;
(ii) a test for chromosomal damage in mammalian cells—cytogenetics in Chinese hamster ovary cells.

In addition, where human exposure might be expected to be extensive and/or sustained and difficult to avoid, a third test for gene mutation in mammalian cells is recommended. Should mutagenic activity be demonstrated *reproducibly* in any *in vitro* test, it would be necessary to investigate whether this potential was also expressed *in vivo* by moving to Stage 2 and, possibly, Stage 3 of the testing strategy. The strategy suggested should not be followed inflexibly, more as a guide. Each substance should be considered on its own and due account taken of chemical structure (any knowledge of its metabolites), toxicokinetics (absorption, metabolism and distribution) and its intended use and hence possible resulting exposure profile for human beings.

5.3 Carcinogens

Of the 100 000 chemicals in common use in daily life, fewer than 30 are known to have caused human cancers, if one excludes certain cytotoxic drugs. Chemicals which induce cancer are called carcinogens.

There are many chemicals which have been shown to cause cancer when dosed to animals, usually rodents. However, it should not be assumed that these experimental or animal carcinogens are necessarily less hazardous to humans than the 30 or so accepted human carcinogens. Many of these latter, human carcinogens were identified in the industrial situations of earlier decades when working conditions were often extremely poor by today's standards. These animal carcinogens may possibly be important workplace hazards which have yet to be detected, either because the added risk is small compared with that due to other causes, or because only a few people have been exposed, or simply because a hazard was not previously recognised and, so far, no workplace studies have yet looked for it. We should also bear in mind that human cancers seldom develop until 10–20 years following exposure to a chemical and it may be too early to detect a chemical carcinogen which may only have been introduced into industry within the last 10 years.

Cancer, or neoplasia (meaning new growth), is a collective term for a particular disease state which can occur in any part of the body. These growths may be either malignant or benign. With the malignant form, the growth usually arises within a particular tissue or organ which continues to increase in size, invades adjacent tissues and often spreads to distant parts of the body. It is this latter property that helps to

STAGE ONE	IN VITRO TESTS
Initial Screening	(a) Bacterial assay for gene mutation.
Two tests (a) + (b) required, except where human exposure would be expected to be extensive and/or sustained and difficult to avoid—when all three tests are necessary.	(b) Clastogenicity assay in mammalian cells.
	(c) Gene mutation in mammalian cells.
STAGE TWO	IN VIVO TESTS
Tests for: Compounds positive in one or more tests in Stage 1 and all compounds where high or moderate prolonged levels of human exposure are anticipated.	(a) Bone marrow assay for chromosome damage.
	Plus, if above negative, and any *in vitro* test positive
	(b) Test(s) to examine mutagenicity or DNA damage in other organs (liver, gut, etc.).
STAGE THREE	IN VIVO TESTS FOR GERM CELL EFFECTS
If risk assessment for germ cell effects is justified on the basis of e.g. pharmacokinetics, use and anticipated exposure.	(a) Interaction with DNA.
	(b) Potential for inherited effects.
	(c) Quantitative* assessment of heritable effects

Fig. 4. Flow diagram for testing strategy for investigating mutagenic properties of a substance (adapted from [9]. *Strong justification required in view of complexity, long duration, costs and numbers of animals required.

distinguish the malignant form from the benign form and, as a consequence, these malignant cancers are often fatal unless surgical removal or other therapies are successful.

It would appear that cancer may be the result of the complex interaction of a multitude of factors, including exposure to viruses, radiation, diet, chemicals and the individual's lifestyle (e.g. smoker *vs.* non-smoker; alcohol; reproductive and sexual behaviour). Moreover, evidence from the USA indicates that the overall cancer death rates are stationary or decreasing, the major exception being smoking-related cancers, particularly lung cancer. The specific cause(s) of many common human cancers have not been identified but dietary factors are thought to be particularly important in the development of cancers of the stomach, colon and rectum.

The total number of cancer cases caused by exposure to chemicals in the workplace every year is not known and estimates vary. The Royal Society put the figure at 5% of total cancer cases and this is comparable to the 4% ascribed by Doll and Peto in 1981 for US cancer deaths attributable to occupational causes [10].

Thus, exposure to carcinogens is not new and is not confined to the workplace. Nevertheless, exposure to certain carcinogens in the workplace, either as individual chemicals or as part of a complex process, has been shown, without doubt, to cause human cancer. For example, vinyl chloride monomer causes liver cancer. The particular liver cancer produced (angiosarcoma—from blood vessel lining cells) is extremely rare in the human population. It was this fact that aided clinicians in the identification of the workplace agent responsible for the increased incidence of the disease.

The development of cancer is thought to involve a number of independent events. The process—carcinogenesis—can be divided into two principal stages: initiation and promotion.

Initiation is thought to involve a somatic mutation in a small number of cells and may be produced by a single exposure to a carcinogen providing this interaction results in the carcinogen reaching the DNA in sufficient quantity to effect initiation.

Initiated cells do not generally develop into cancer cells unless subjected to a prolonged promotion phase, either by further carcinogen exposure or by contact with a promoting agent which may be unable to induce cancer on its own. The period between initial exposure to a chemical carcinogen and the occurrence of cancer, which may be as long as 40 years in man, is called the **latent period**.

The majority of chemicals which induce the development of cancer in humans have been demonstrated to induce tumours in experimental animals. Lifetime experiments in rodents (rats and mice) are the classic means of detecting carcinogens.

The decision to undertake a full carcinogenicity study requires justification since they are extremely costly, involve a large number of animals and tie up skilled personnel and valuable resources for a long time. Nevertheless, because of the severity of the harmful outcome, there are circumstances in which the use of these studies is appropriate. In general, carcinogenicity studies are required:

(i) when it is anticipated that the use of the chemical is likely to result in human exposure over a substantial period of life (6 months or more),
(ii) where a chemical (or a metabolite) is structurally related to a known carcinogen,
(iii) where some aspect of the biological activity of the chemical gives cause for concern,
(iv) the toxicity profile from sub-chronic investigations of the chemical are indicative of pre-neoplasia (such as hyperplasia or dysplasia) and/or there is evidence of long-term retention of the chemical (or its metabolites) within the body or
(v) there are positive results from one or more of the screening tests for mutagenic or potential carcinogenic activity.

The experimental protocol should take account of regulatory requirements, scientific concerns and the animals' welfare. A number of bodies have suggested guidance for the conduct of carcinogenicity studies and, nowadays, the basic features are broadly agreed. As far as industrial chemicals are concerned, for carcinogenicity studies, the

Good Laboratory Practice (GLP) guidelines [11] in conjunction with protocols produced by the OECD and the United States Environmental Protection Agency (EPA) are normally utilized [12,13].

The results of a carcinogenicity study are based on a comparison of the nature, incidence and time of occurrence of tumours in treated animals when reviewed alongside that of a concurrent untreated or control group of animals. In order to state that any chemical is positive in a carcinogenicity study, one has to demonstrate that the test chemical has induced:

(a) an increased incidence of tumours,
(b) an earlier appearance of these tumours, or
(c) the incidence of an unusual tumour type,

any one of which (a, b, or c) must be judged in comparison with spontaneous tumours found in the concurrent control group.

A number of major scientific and technical factors should be considered when planning carcinogenicity studies, most of which have been the subject of intensive debate in recent years. In general, these concerns relate to one of three areas—the test animal, the test material and the test procedure. Some of these issues are outlined in Table 3. Whilst it is not the intention to detail each and every concern highlighted in Table 3, it is worth describing, in general terms, the following important points.

Table 3. Particular points of concern in relation to the design of carcinogenicity studies

---

The test animal
Species
Strain
Sex
Size of experimental group
Knowledge of background incidence of spontaneous tumours

The test material
Analytical specification—fingerprint
Technical grade versus 'purified' sample
Stability in vehicle used for administration

The test procedure
Inception and duration of the test
Route of exposure
Dose level and frequency of exposure
Clinical observations during the study
Post-mortem procedures
Histopathological evaluations
Statistical considerations: randomization of animals initially and analysis of terminal data

---

Test animal
Experience coupled with practicality and availability usually determine that conventional carcinogenicity tests are done in rats and mice. The selection of strain is more difficult and usually the toxicologist will decide in favour of well-validated strains, which may be in-bred or out-bred, and for which there is detailed information relating to the incidence of spontaneously occurring tumours together with a low incidence of non-tumorigenic diseases. The experimental group size usually consists of 50 male and 50 female animals in order that any sex differences in responses to a chemical may be investigated. This also allows sufficient numbers for statistical analyses of the results where any increased incidence in tumours may be masked by a low spontaneous incidence of the same tumour type in the control animals.

Test material
In selecting chemicals for carcinogenicity testing, it is essential to have basic information on the physico-chemical parameters of test material, such as its stability in water and other solvents, which may be used during the dosing procedures, together with its impurities. Most industrial chemicals, commonly termed 'technical grade', contain impurities. Since human beings are exposed to this grade of material in the workplace, these are used routinely in carcinogenicity studies in the first instance. Since several different batches of a test material may be required during a two year study involving three dose levels, it is imperative that the investigator knows the degree of comparability of these batches with respect to their composition and stability.

Test procedure
Conventional carcinogenicity studies begin by dosing shortly after weaning, when rodents are between the ages of 5–7 weeks, and usually continue for 2 years in the rat and 1.5 years in the mouse. One objective of carcinogenicity testing is to obtain the maximum possible carcinogenic effect of a test substance in order to compensate for the limited numbers of individuals at risk. This is accomplished by employing high levels of exposure and by starting the treatment when the animals are at their most sensitive.

The extent of dosing or exposure is a compromise, chosen on empirical grounds to balance a number of factors such as:

(i) spontaneous tumours usually occur with increasing frequency late in life and their appearance may make it more difficult to evaluate the carcinogenicity of a test material, especially if the test material is of low potency;
(ii) the wish to maximize the detection of an oncogenic effect by continuing treatment for a longer period.

A diagrammatic representation of the main elements for a carcinogenicity protocol are illustrated in Table 4.

Animals should be treated by the likely route of exposure for man, at least insofar as oral, dermal and inhalation exposures are concerned. For most industrial chemicals, the oral route is routinely used.

Table 4. Example of a rodent carcinogenicity protocol for an industrial chemical (based on OECD/EPA[a] guidelines)

Protocol component	Factor
Study management	
Route of exposure	Most usually oral—diet/gavage
Frequency of exposure	Most usually 5 days/week (7 days preferred)
Duration of exposure	Mouse: 18–24 months. Rat: 24–30 months
No. of treatment groups (T)	3
No. of control groups (C)	1
No. of animals per group (T or C)	At least 50 animals of each sex
Age at start of study	Usually 5–7 weeks
No. of animals per cage	Single, if available, or up to 4 per cage depending on size of cage
Photoperiod	12 hours day and 12 hours night
Temperature	22°C (plus or minus 3°C)
Humidity	30–70%
Air changes per hour	at least 15
Study observations:	
Body-weights	1/week for first 13 weeks; 1/month after
Food consumption	1/week for first 13 weeks; 1/month after
Clinical signs and intercurrent deaths	At least once per day, preferably twice
Haematology	Differential blood count at 12, 18 months and termination for high dose group and control—other groups if indicated
Post-mortem and histopathology:	
Autopsy	All animals
Extent of tissues required routinely	Adrenals, Bladder, Bone (sternum), Brain, Colon, Duodenum, Eyes, Femur (+marrow), any 'gross lesion', Heart, Ileum, Jejunum, Kidneys, Liver, Lungs, Mammary gland, Mesenteric lymph node, Oesophagus, Pancreas, Pituitary, Prostate/uterus, Salivary gland, Spinal cord (three levels), Spleen, Stomach, Submandibular lymph node, Testes/ovary, Thymus, Thyroid/parathyroid, Trachea
Additional tissues required occasionally	Blood smear, Caecum, Epididymis, Femur (including joint), Muscle, Peripheral nerve, Rectum, Seminal vesicle, Skin, Vagina

[a] EPA, Environmental Protein Agency (of the USA).

Routinely conducted carcinogenicity studies (such as those conducted under the auspices of the National Toxicology Program, NTP in the USA) have only two treatment groups. It is, however, advisable to have three treatment groups (low, intermediate and high dose) to enable the construction of a dose–response curve and an investigation of this relationship and the promulgation of a 'no observed effect level'. The general aim, when deciding on dose levels for administration in a carcinogenicity assay, is to maximize exposure. However, this must be tempered by a consideration of the toxicokinetics and toxicity of the chemical under investigation. Usually, the high dose level is set at the maximum tolerated dose (MTD), the low dose at a small multiple of the likely human exposure, and the intermediate dose at the mean of these two values. If the test substance does not have any specific target organ toxicity (identified from sub-chronic testing, e.g. a 90-day study), the high dose is usually set at a level which is likely to cause some 10% reduction in body-weight gain over the animal's lifespan. A further constraint on this top dose is that it should not normally exceed 5% of the weight of the diet or drinking water if administered in either of these vehicles.

The basic purpose of the carcinogenicity study is to provide details relating to tumour incidence in treated animals compared with the control group. Thus, it is essential that routine daily checks on the animal's health and general behaviour are made and recorded, together with any signs of development of visible masses. The reader is referred to OECD Test Guideline 451 for full details of the study requirements.

Clearly, the control and analysis of the data generated in these types of chronic investigation must be extremely thorough and exhaustive so that the possibility of false negatives is eliminated. Interpretation of the results of carcinogenicity studies is fraught with difficulties, even with well-conducted studies, especially when the results of such an animal investigation are at variance with well conducted studies of human populations (epidemiology). An increase in the incidence of tumours in test groups compared with that in the control group is not sufficient to attribute causally the increased incidence to the test chemical. A number of factors need to be considered before any such causal link can be established; these include:

— the dose–response relationship, if any,
— the nature and type of tumour induced, and
— other factors affecting tumour incidence—genetic, dietary, hormonal, viruses or physico-chemical effects.

The majority of chemicals which cause human cancers have been demonstrated to induce tumours in experimental animals. These lifetime studies in rodents are the classical means of detecting carcinogens. Nevertheless, there are some disadvantages in these assays relating to the high doses of chemical administered. Some authorities suggest that these 'megadoses' overwhelm the normal pathways of metabolism or detoxification. Hence, extrapolation of results from animals to humans, which is difficult anyway, is unlikely to be appropriate given the abnormal biochemical and physiological processes pertaining in this 'overload' scenario.

Despite these limitations, the similarities in the anatomy and physiology of mammalian species (including rodents and humans) makes possible the prediction of toxic effects of chemicals in humans.

Chemicals which give a positive result in an animal carcinogenicity study are given a higher priority for the introduction of measures to control exposure in the workplace.

5.4 Teratogens

There are a limited number of examples of the effects of exposure of industrial chemicals on the reproductive capacity of both men and women, as well as the reproductive outcome. Indeed, there is very little human evidence of harmful reproductive outcome, whether it be loss of libido or overt foetal abnormalities. The thalidomide experience was undoubtedly the most significant event in influencing the rationale behind reproductive toxicity evaluation. This tragedy led legislators to attempt to strengthen or develop new codes of practice for investigating the potential of chemicals to cause malformations or birth defects. This understandable concern has meant that investigation of birth defects has overshadowed the investigation of other adverse or potentially adverse effects of agents upon reproduction. This is despite the fact that overt birth defects are probably among the least likely hazards to reproduction of a compound. Nevertheless, one has to acknowledge the public, political and commercial concerns relating to this particular aspect of reproductive toxicity and only with the Seventh Amendment will there be a widening of the classification relating to adverse reproductive outcome. With this revision of the CPL of dangerous substances, there will be a new category—Toxic to Reproduction—which will allow for the classification and labelling of substances which impair either structural or functional capacity in humans. Thus, teratogenic effects will be correctly seen as one of a spectrum of possible deficits of reproductive capacity induced by exposure to external agents.

The mammalian reproductive cycle is a complex process, involving the following events, each exhibiting its own peak of sensitivity to extraneous factors:

Gametogenesis: development of male (spermatozoa) and female (oocytes) germ cells
Mating
Fertilization
Blastogenesis: formation of the blastocyst
Implantation: attachment of blastocyst to uterus wall
Embryogenesis: differentiation of embryonic cells into various organ systems
Foetogenesis: period of general growth, organ differentiation, external genitalia and
         histogenesis of central nervous system
Gestation period: from fertilization to parturition
Parturition: birth of the young
Prenatal period: between fertilization and parturition
Perinatal period: shortly before and after parturition
Post-natal period: follows parturition
Lactation period: first part of post-natal life from parturition up to weaning of the
         offspring.

In order to study the effects of extraneous factors on this complex process, the cycle is conveniently divided into the following periods:

(i) the formation of the gametes,
(ii) gestation, and
(iii) post-natal life.

The effects of an industrial chemical on all these aspects of the reproductive cycle are investigated using a battery of tests which includes:

(a) the multi-generation study,
(b) the teratogenicity study, and
(c) the peri- and post-natal study.

Multi-generation study—this is an overall screening study, covering the entire reproductive cycle in two or more successive generations of animals, exposure to the test substance being continuous throughout the study. The majority of these studies use the rat as the model. A variant of this study—the fertility and general reproductive performance study—is a specialized assay where the test substance is administered directly to the first (parent) generation only. However, the reproductive ability of the offspring of this first generation is also assessed.

Teratogenicity study—this study concentrates on the most sensitive part of gestation, from the time of implantation until major organogenesis is complete. This period represents the time when a test substance, administered to the pregnant female, is most likely to induce malformations in the embryo. Regulatory guidelines for these assays suggest that studies should be conducted in two species, one in a rodent, usually the rat, and the second in a non-rodent species, the rabbit.

Peri- and post-natal study—this concentrates on the latter part of gestation, not covered by the teratogenicity study, and which encompasses parturition and the lactational period. This study may be particularly useful in the detection of subtle effects on the brain, which is not fully developed, either physically or functionally, during the dosing regime of a teratology study. The rat is the preferred model for this study.

The three study designs above are often referred to as Segment I, Segment II and Segment III studies, respectively. In general, it is possible initially to limit reproductive toxicity testing to two key areas, namely fertility and teratogenicity.

5.5 Toxicokinetic investigations

The term toxicokinetics is one applied to the study of the absorption, distribution, metabolism and excretion of non-pharmaceutical products as a function of time in the body. The equivalent term 'pharmacokinetics' is used for drugs and pharmaceuticals.

A number of scientific groups, such as the WHO and the EC Scientific Committee for Food (SCF), have stated that toxicokinetic studies should be undertaken as soon as the acute toxicity of a chemical has been determined. This enables scientists to understand mechanisms, dose-rate effects and the formation of active metabolites from industrial chemicals. These latter factors are important when they may indicate differences in metabolic handling between species which may be chosen for chronic toxicity studies.

Ideally, toxicokinetic studies should be carried out on all chemicals prior to any chronic study to aid our understanding of toxic mechanisms and bioavailability and to help in the final assessment of the results of these chronic studies when one has to evaluate the risk to man. Unfortunately, in most cases, these investigations are hard to justify in the simple determination of the presence or absence of toxic effects.

6. RISK ASSESSMENT

6.1 Introduction

The basic concept of experimental toxicology is the identification of the intrinsic biological properties of chemicals. Chemicals exhibit a spectrum of effects as a (result of these intrinsic properties) and whether they are beneficial or detrimental to human health and the environment is ultimately related to the dose. The procedures for determining the potential benefits and/or adverse side effects from exposure to chemicals form the basis of risk assessment.

Risk assessment may be defined as a process of decision making when there may be a number of possible outcomes and it is uncertain which event will occur. Risk assessment can be broken down into a number of stages which may be summarized as follows (see earlier sections for definitions):

(a) hazard identification,
(b) hazard evaluation,
(c) exposure evaluation, and
(d) risk estimation.

The preceding sections have effectively covered the first two stages of this process. A review of the results of toxicity testing, usually in animals, together with a determination of such factors as dose–response, species variability in mode of action and susceptibility, and the toxicokinetics of the chemical should form the basis of the hazard identification and evaluation steps. The third step, the assessment and evaluation of exposure in individuals or populations, is a prerequisite of the quantification of risk. This evaluation procedure may be based upon an estimation, via modelling, of the probable human exposure taking into account various routes of entry and the levels of likely exposure.

The estimation of human exposure to an industrial chemical within UK industry since the introduction of the COSHH Regulations and its requirements for the keeping of records of personal exposure to chemicals 'hazardous to health' will hopefully allow more realistic evaluations to be made than in past decades. Then much information was not kept and/or monitoring of the workplace environment was not routinely performed, as personal sampling on the operator, rather than static samples, were the norm.

The two main approaches to estimation of the risk to humans from exposure to industrial chemicals are generally referred to as:

— the safety factor approach, and
— mathematical modelling.

6.2 Safety factor approach

This method was developed primarily with a view to providing a legal framework for the control of certain additives in foodstuffs and was promoted by the Americans. However, it has found wide application in the setting of 'threshold limit values' for industrial exposures.

The method relies upon the determination of an acceptable daily intake (ADI) for a chemical. The ADI is derived by division of the NOEL, the no observed effect level, by a *safety factor* (SF).

$$\text{ADI} = \frac{\text{NOEL}}{\text{SF}}$$

The SF is traditionally set at 100 but, in some cases, may be as low as 10. This level is an empirical figure but is stated to reflect the possibility that humans may be 10 times more sensitive than a particular test species, in any study from which the NOEL was derived, and also to allow for a 10-fold range in susceptibility within any human population which might be exposed to the chemical in question.

This approach to the estimation of risks has been criticized on the grounds that:

(a) the NOEL is highly dependent upon the number of animals in a study group—the smaller the number of animals, the higher the NOEL is likely to be,
(b) that it does not take account of the slope of any dose–response curve and
(c) that the NOEL appears to support the concept of a 'threshold dose' below which a chemical does not induce toxic effects—this may not be true for all carcinogens.

Nevertheless, despite these deficiencies, this approach is still widely utilized and it does incorporate the estimation of both low dose effects and inter-species relationships in one method.

6.3 Mathematical modelling

A number of models have been developed in an attempt to describe the variations in dose–response data seen in toxicological studies. All these models aim to extrapolate from the relatively high exposure levels seen in animal studies (or from values measured in occupational surveys) to determine the risk(s) to the general population from low exposure levels in the ambient environment.

If we base this section on the risk of cancer following exposure to a genotoxic carcinogen, it is currently accepted that any exposure, however small, may lead eventually to an increased risk of cancer. The major problem with any risk assessment relates to the choice of the most appropriate model. The most commonly utilized models in cancer risk assessment are shown in Table 5. By reference to Fig. 5, the

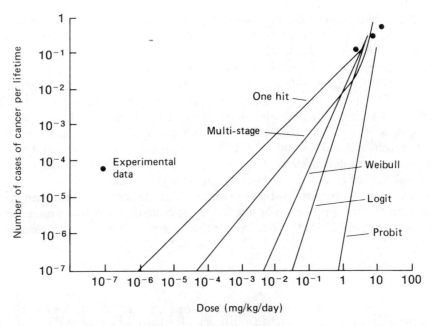

Fig. 5. Relationship between cancer risk and dose and dependence upon mathematical model.

Table 5. Some models used in cancer risk assessment

Category	Model
Mechanistic (stochastic)	One hit Multi-stage Linearized multi-stage
Empirical	Weibull Logit Probit

reader will clearly see that the risks described by this method of mathematical modelling are highly dependent upon which model is chosen [14]. In the USA, the regulatory authorities use the linearized multi-stage model in which risk is assumed to be directly proportional to the dose, even at low dose levels. This model is one of the most 'conservative' and application of this model to toxicological data from animal studies would tend to provide high estimates of risk. It is a model which errs on the side of caution. With respect to exposure to carcinogenic chemicals, most regulatory bodies define an 'acceptable risk' as one where the daily exposure would increase the incidence of cancer from around one person in 10^7 to one in 10^5. Risks of this magnitude, which may induce a certain cancer incidence, cannot be detected by studies of exposed human populations (epidemiology) and are thus referred to as virtually safe.

All these mathematical models make a number of assumptions since they are not able to take account of the complexity of events which occur following initial exposure to a chemical through to the development of a tumour, not to mention the toxicokinetic and possible 'repair processes' innate in human defence mechanisms. Thus their usefulness is limited. Wherever possible, human data should be factored into the overall assessment of any risk rather than too much reliance be placed on these models which are based on animal studies and a number of unproven and perhaps scientifically naive assumptions.

The full appreciation of all available toxicological data, be it qualitative or quantitative, in combination with the wise use of mathematical models, will enable a more satisfactory approach to be applied to the evaluation of risks to man and the promulgation of scientifically defensible controls of exposure, leading to greater safety for all.

7. CONCLUSIONS

The public's heightened awareness of three broad areas of adverse health effects—cancer, neurological impairment and birth defects—is based upon the fact that these conditions are irreversible. This increased concern with respect to the potential of chemicals to cause adverse human health or environmental effects has been influenced, in part, by the publicity following accidents at chemical plants, such as, Flixborough, Seveso and Bhopal. This has been augmented by the attendant press coverage arising from the side-effects of various medical treatments, such as Opren and Thalidomide.

Since Doll and Peto's report [10] that only 4% of cancer cases were of occupational origin, the public have realized the implication: in excess of 80% of all cancers are environmental in origin. This analogy alone has added to the public's concern and focused their attention on the chemical additions to the environment.

The public, quite correctly, demand that regulatory authorities should protect them from chemicals and other agents which may cause these adverse effects by way of specific control measures. They want reassurances of 'safety' in going about their daily lives which may bring them into contact with numerous chemicals and agents which may induce adverse health or environmental effects at specific concentrations or doses. However, some regulatory controls may be diluted unless the public are prepared to help themselves. For example, there is an abundance of knowledge about some of the causes of certain site-specific cancers and heart disease. Lowering the intake of total fat in the diet, ingesting more dietary fibre and refraining from smoking would significantly reduce the incidence of a great many human cancers. Regulatory actions alone will not achieve a fraction of the reduction in cancer incidence that could be achieved by a few simple alterations in lifestyle.

Toxicology is only useful in health issues if its knowledge is translated meaningfully and realistically into public actions.

Over the past 20–30 years, much progress has been made in assessing the risks of adverse health effects from chronic, low-level exposure to chemicals. Classically, this was done by epidemiological studies, often in the workplace, or by long-term studies in rodents. Nowadays, other more economic and rapid assays can form a 'test battery' which can reliably predict possible carcinogenic effects. In the future, it is anticipated that further advances in molecular biology will aid in the understanding of the mechanistic basis of these chronic health effects and hence improve this risk assessment process.

Confidence in the ability of regulators and governments to distinguish risks would be enhanced if the public were better informed. A consensus in the extrapolation of data from animal studies to man of a potential human health risk would be more acceptable if the public were convinced that the primary purpose of governmental control was not merely to punish offenders, or permit a degree of ill-health, but to achieve some rational basis whereby the positive aspects of a chemical might be retained whilst protecting the public health. Legislation which encourages the improvement of risk assessment and programmes which aim to reduce the uncertainties inherent in these assessments are the best long-term solutions for addressing any determination of chemical safety. These decisions should be made in the light of scientific consensus and the best science available at the time.

There is no such thing as 'absolute safety'. All activities involve some degree of risk. It is therefore necessary to decide the level of risk which is acceptable and what priorities/controls should determine how this risk is to be managed. Society is becoming ever more aware of the issues and thus it is important that any decisions relating to the control of chemicals which have an effect on the environment or health are assessed as openly as possible.

REFERENCES

[1] W. N. Aldridge, 1981, *Trends Pharmacol. Sci.* **2**, 228–231.
[2] H. P. A. Illing, 1991, *Ann. Occup. Hyg.* **35**(6), 569–580.
[3] 92/32/EEC, 'Seventh Amendment Directive', *Official Journal of the European Communities*, L 154, **35**, 5.6.92, HMSO.
[4] A guide to the Notification of New Substances Regulations, HS(R)14 (Rev), 1988.
[5] ACGP, Annex to Commission Directive 84/449/EEC, Official Journal, of the European Community, L251, Vol. 27, dated 19.9.84, HMSO.
[6] OECD Guidelines for Testing of Chemicals: Sect. 1–IV. Physical-chemical properties; Effects on biotic systems; Degradation and accumulation; and Health effects, 1979, OECD.
[7] Test Methods of Annex V to Council Directive 79/831/EEC.
[8] Regulation on the Evaluation and the Control of the Environmental Risks of Existing Substances (Doc.XI/170/89), Official Journal of the European Communities, December 1989, HMSO.
[9] 'Guidelines for the Testing of Chemicals for Mutagenicity', Department of Health Report No.35, 1989, HMSO.
[10] *The Causes of Cancer* R. Doll and J. Peto 1981, Oxford University Press.
[11] Good Laboratory Practice. Home Office Code of Practice for the Housing and Care of Animals used in Scientific Procedures, 1989, HMSO.

[12] OECD Guidelines for Testing of Chemicals: Carcinogenicity Testing, Number 451, 1981, OECD.
[13] Environmental Protection Agency (EPA) Health Effects Testing Guidelines—Chronic Exposure 40 CFR 798, 50 FR 39252, 27.9.85, HMSO.
[14] 'Guidelines for the Evaluation of Chemicals for Carcinogenicity', Department of Health Report No. 42, 1991, HMSO.

Appendix I: Glossary of acronyms

ACGIH	American Conference of Governmental Industrial Hygienists
ACoPs	Approved Codes of Practice (adopted by the UK Health and Safety Commission)
ACTS	Advisory Committee on Toxic Substances (to the HSC)
ADs	Appointed Doctors
ADIs	Acceptable Daily Intakes
AFOM	Associate of the Faculty of Occupational Medicine (at the RCP)
AHH	Aryl Hydrocarbon Hydroxylase (induction test for chloracne)
AIHC	American Industrial Health Council
AM	arithmetic mean
ANSI	American National Standards Institute
APME	Association of Plastics Manufacturers in Europe
ARAB	Algemeen Reglement voor de Arbeidsbescherming (General Regulations for the Protection of Labour, Belgium)
Arbo Act	Arbeids Omstandinghedenwet (Working Environment Act, The Netherlands)
BAA	British Agrochemicals Association
BATs	Biologische Arbeitsstofftoleranzwerte (Biological Tolerance Values for Working Materials, Germany)
BEIs	Biological Exposure Indices (adopted by the ACGIH)
BELs	Biological Exposure Limits
BGs	Berufsgenossenschaften (employers' associations for accident insurance, Germany)
BG Chemie	the BG for the chemical industry (Germany)
BIAC	Business and Industry Advisory Committee
BLS	Bureau of Labor Statistics (USA)

Appendix I

BLVs	Binding Limit Values (EC)
BOHS	British Occupational Hygiene Society
BP	British Pharmacopoeia
BRMA	British Rubber Manufacturers' Association
BSI	British Standards Institution
CEC	Commission of the European Communities
CEFIC	European Chemical Industry Council
CEN	Comité Européen de Normalisation (European Committee for Standardization)
CFCs	chlorofluorocarbons
CFR	Code of Federal Regulations (USA)
CIA	Chemical Industries Association
CIIT	Chemical Industry Institute of Toxicology (USA)
CISHEC	Chemical Industry Safety, Health and Environment Council (of the CIA)
CMA	Chemical Manufacturers' Association (USA)
CNS	central nervous system
CONCAWE	Oil Companies' European Organization for Environmental Health Protection
COREPER	Committee of Permanent Representatives (to the EC)
COSHH	Control of Substances Hazardous to Health (Regulations, 1988)
CPC	chemical protective clothing
CPL	Classification, Packaging and Labelling (Regulations, 1984)
CRAM	Caisses Régionales d'Assurance des Malades (Regional Sickness Insurance Funds, France)
DFG	Deutsch Forschungsgemeinschaft (the German Research Society)
DIN	Deutsche Industrie Normung
DNA	deoxyribonucleic acid
DSS	Department of Social Security (of the UK Government)
EC	European Community
ECETOC	European Centre for Ecotoxicology and Toxicology of Chemicals
EH 40	HSE guidance booklet on Occupational Exposure Limits (revised annually)
EH 42	HSE guidance booklet on Monitoring Strategies for Toxic Substances
EHC	Environmental Health Criteria (documents, produced by the IPCS)
EINECS	European Inventory of Existing Commercial Chemical Substances
ELINCS	European List of Notified (New) Chemical Substances
EMAs	Employment Medical Advisers
EMAS	Employment Medical Advisory Service
ENs	Enrolled Nures
EP	European Parliament (of the EC)
EPA	Environmental Protection Agency (USA)
ERL	Environmental Resources Limited (UK consultants)

ESC	Economic and Social Committee (of the EC)
ETSs	Emergency Temporary Standards (set by OSHA, USA)
EUROMETAUX	European Association of Metals
FAO	Food and Agriculture Organization (of the UNO)
FFOM	Fellowshoip of the Faculty of Occupational Medicine (at the RCP)
GIFAP	Groupement International des Associations Nationales de Fabricants de Produits Agrochimiques (International Group of National Associations of Manufacturers of Agricultural Products)
GLP	Good Laboratory Practice
GM	geometric mean
GMC	General Medical Council
GPs	general practitioners
GSD	geometric standard deviation
HAZOPs	Hazard and Operability Studies
HEDSET	Harmonized Electronic Data Set (EC/OECD)
HMFI	Her Majesty's Factory Inspectorate
HPV	high production volume (chemicals)
HSC	Health and Safety Commission (of the UK)
HSE	Health and Safety Executive (of the UK)
HSGs	Health and Safety Guides (produced by the IPCS)
HSW Act	Health and Safety at Work etc. Act 1974
IARC	International Agency for Research on Cancer (of the WHO)
ICCA	International Council of Chemical Associations
ICOH	International Commission on Occupational Health
ICSCs	International Chemical Safety Cards (issued by the IPCS)
IgE	immunoglobulin E
IIAC	Industrial Injuries Advisory Council (of the UK Government's DSS)
ILO	International Labour Organization (Note: The International Labour Office is the secretariat of the International Labour Organization)
ILVs	Indicative Limit Values (EC)
IM	intramuscular
IMO	International Maritime Organization
IOSH	Institution of Occupational Safety and Health
IPCS	International Programme on Chemical Safety
IRPTC	International Registry of Potentially Toxic Chemicals (maintained by UNEP)
ISO	International Standards Organization
IV	intravenous
JACC	Joint Assessments of Commodity Chemicals (reports produced by ECETOC)
JECFA	Joint WHO/FAO Expert Committee on Food Additives
JMPR	Joint Meeting on Pesticide Residues

LC 50	Lethal Concentration (in air) 50%
LD 50	Lethal Dose 50%
LEV	Local Exhaust Ventilation
MACs	Maximum Acceptable Concentrations (adopted in the former Soviet Union)
MAK Commission	Maximale Arbeitsplatzkonzentrationen (MAK) Commission (Germany)
MAK Values	Maximale Arbeitsplatzkonzentrationen (MAK) Values (Maximum Concentration Values in the Workplace, Germany)
MDHS	Methods for the Determination of Hazardous Substances
MELs	Maximum Exposure Limits
MFOM	Member of the Faculty of Occupational Medicine (at the RCP)
MTD	maximum tolerated dose
NADOR	Notification of Accidents and Dangerous Occurrences Regulations (1980)
NAMAS	National Measurement Accreditation Service
NIOSH	National Institute for Occupational Safety and Health (USAa)
NOEL	no observed effect level
NONS	Notification of New Substances (Regulations 1982)
NPF	Nominal Protection Factor
NTP	National Toxicology Program (USA)
OECD	Organization for Economic Co-operation and Development
OELs	Occupational Exposure Limits
OESs	Occupational Exposure Standards
OH	Occupational Health
OHC	Occupational Health Centre
OHNC	Occupational Health Nursing Certificate (for RGNs)
OHND	Occupational Health Nursing Diploma (for RGNs)
OHS	Occupational Health Service
OJ	*Official Journal of the European Communities*
OSHA	Occupational Safety and Health Administration (of the USA)
OSHAct	Occupational Safety and Health Act (1970, of the USA)
PCIAOH	Permanent Commission and International Association on Occupational Health
PELs	Permitted Exposure Limits (USA)
pH	minus the logarithm (to base 10) of the hydrogen ion concentration (a measure of acidity or alkalinity of aqueous solutions)
PPE	personal protective equipment
QMV	Qualified Majority Voting (in the EC)
R 45	EC labelling risk phrase, 'May cause cancer'
R 49	EC labelling risk phrase, 'May cause cancer by inhalation'
RAST	Radioallergosorbent (test)
RCN	Royal College of Nursing
RCP	Royal College of Physicians
RGN	Registered General Nurse

RGPT	Règlement Général pour la Protection du Travail (General Regulations for the Protection of Labour, Belgium)
RICE	Regular Inter-laboratory Counting Exchange
RIDDOR	Reporting of Injuries, Diseases and Dangerous Occurrences Regulations (1985)
RPE	respiratory protective equipment
RSC	Royal Society of Chemistry
SCBA	self-contained breathing apparatus
SCF	Scientific Committee for Food
SEA	Single European Act (1986)
SF	safety factor
SIDSs	Screening Information Data Sets (of the OECD Existing Chemicals Programme)
SOPs	Standard Operating Procedures
STELs	short-term exposure limits
TLVs	Threshold Limit Values (registered trademark, adopted by the ACGIH)
TPCs	Technical Progress Committees (EC)
TQM	Total Quality Management
TRK Values	Technische Richtkonzentrationen (Technical Exposure Limits, Germany)
TWA	time-weighted average
UNEP	United Nations Environment Programme
UNICE	Union de Confédérations de l'Industrie et des Employeurs d'Europe (Union of Industrial and Employers' Confederations of Europe)
UNO	United Nations Organization
WASP	Workplace Analysis Scheme for Proficiency
WATCH	Working Group for the Assessment of Toxic Chemicals (reporting to ACTS)
WHO	World Health Organization
WTR	International Working Group on the Toxicology of Rubber Additives

Appendix II: Some health publications issued by CIA

A. GUIDANCE BOOKLETS ON THE CONTROL OF SUBSTANCES HAZARDOUS TO HEALTH (COSHH) REGULATIONS

—Collection and Evaluation of Hazard Information (Regulation 2)
Identifying hazardous substances and the nature of the health hazards they present are important tasks in assessing the risks to health posed by such substances in the workplace. Guidance is given in this booklet on how to identify such substances and sources of information for assessing the risks they may present.
Published 1991.

—Assessments (Regulation 6)
This booklet is aimed primarily at management in CIA member companies to help them meet the requirements of Regulation 6 of COSHH to carry out assessments. It offers practical guidance rather than a detailed interpretation of the regulation. The principles should also be helpful and relevant to the many industries using chemicals.
Published 1989.

—Setting In-House Occupational Exposure Limits (Regulation 7)
COSHH Regulation 7 requires workplace exposures to substances hazardous to health to be adequately controlled. In respect of substances which have been assigned official exposure limits (i.e. Maximum Exposure Limits or Occupational Exposure Standards), this means compliance with those limits. However, for all other substances, companies are expected to set their own in-house working standards. This booklet

gives practical guidance on how to do this by way of a systematic procedure comprising collection and review of technical data, qualitative assessment of the hazards and quantification of the risks.
Published 1990.

—Exposure Limits for Mixtures (Regulation 7)
This booklet explains that the toxic effects of two or more substances present together in the workplace atmosphere can combine in various ways but, principally, additively, independently or a combination of both. It gives a numerical approach for deriving appropriate limits for such mixtures where limits are known for the components acting alone. Generally, an additive approach can be applied although some mixed systems can be considered by an independent or other approach. Worked examples illustrate the principles involved.
First published 1985; Second Edition 1993.

—Chemical Protective Clothing (Regulation 7)
Consisting of guidance on selection, use and maintenance, information, instruction and training, this booklet concentrates on the dangers of toxic chemicals penetrating the material of protective clothing. It is aimed chiefly at protective gloves, but the advice is also applicable to protective overalls and footwear. A glossary summarizes the principal properties of materials commonly used in the manufacture of chemical protective clothing and gives brief explanations of some associated terms. The Second Edition has brought the originally published guidance up to date with the COSHH Regulations.

It is hoped that this guidance will encourage manufacturers of protective clothing to improve their technical literature on the resistance of their products to chemicals.
First published 1986; second edition 1993.

—Health Surveillance (Regulation 11)
Regulation 11 of COSHH outlines the general principles of health surveillance and provides a framework for decision-making thereon. This booklet provides guidance to help employers decide:

—whether health surveillance is required,
—if so, what kind of surveillance,
—who should carry it out, and
—what records should be kept.

Although intended primarily for management in CIA member companies, it should prove useful and relevant to companies in the many industries using chemicals.
Published 1989.

—Information, Instruction and Training (Regulation 12)
COSHH Regulation 12 requires companies to provide their employees who may be exposed to substances hazardous to health with 'suitable and sufficient' information, instruction and training on the health risks and appropriate precautions to be taken.

Specifically, this includes the results of exposure monitoring—which may be carried out in accordance with COSHH Regulation 10—and the collective results of health surveillance performed in accordance with Regulation 11.

Practical guidance on these requirements is given in this booklet in the following terms:

Information: facts and their interpretation relating to hazards, risks and precautionary measures;
Instruction: showing people what to do in respect of controls, precautions, emergency and other procedures, and
Training: ensuring that information and instruction have been understood and are implemented in practice.

Published 1990.

—Record Keeping

A number of the COSHH regulations require records to be kept in order to promote safe working practices for dealing with hazardous substances. This booklet summarizes both the information which is required to be kept by the regulations and certain other information which may be useful to achieve compliance. It should be read in conjunction with the text of the regulations and the COSHH general Approved Code of Practice.

Published 1991.

—COSHH in distribution

This is intended for distribution personnel in the chemical industry and for their transport and warehouse contractors. It outlines the aspects of the COSHH Regulations which apply to distribution.

Published 1992.

B. OTHER HEALTH PUBLICATIONS

—Allergy to Chemicals at Work

This booklet is aimed at helping line managers and their medical advisers to understand and cope with a series of common health problems at work caused by occupational allergens. The first part outlines general aspects, definitions and principles of good management in relation to the responsibilities of the occupational physician. The second part deals with technical aspects, including the identification of allergens, medical surveillance and clinical identification of occupational allergy together with regulatory controls.

Published 1983.

—Carcinogens in the Workplace

This publication is designed to assist all those who, during their work, may encounter chemicals which can cause cancer. Its purpose is to explain the considerations which should be taken into account in order that reasonable and responsible actions can

be taken. It was produced in response to an important need identified by CIA's National Joint Advisory Committee on Health and Safety in the Chemical Industry on which all trade unions associated with the industry are represented.
First published 1987, Second Edition 1993.

—Chemical Exposure Treatment Cards
This book of 22 cards contains authoritative recommendations for first aid and medical treatment of over-exposure to about 90 of the most commonly encountered gases, vapours and volatile liquids. The cards should be used to ensure that necessary equipment is available wherever such accidental contamination can take place and indicate the skills which may be required by first aiders in the treatment of casualties. They are also used to record the treatment which a casualty may have already been given at a company site and indicate the advice which can be useful in hospital if further attention is needed. Copyright has been waived to allow users to photocopy individual cards.
First published 1989; Second Edition 1991.

—Employment and Reproductive Health
This booklet examines factors of work which can affect the ability of the reproductive systems of both sexes to produce healthy offspring. It reviews the nature and extent of reproductive health problems in industry with an explanation of the stages in the reproductive process and the effects produced by toxic agents. The use of toxicity data from animal experimentation and human investigation is also described. The booklet offers practical guidelines to companies for tackling this complex subject in order to protect employees from hazards to reproductive health.
Published 1984.

—Formaldehyde—Questions and Answers
This introduction to formaldehyde has been written by CIA's Formaldehyde Health Impact Study Team, which is concerned with the chemical and medical hazards connected with the manufacture and use of formaldehyde.

The publication looks at varios aspects of formaldehyde, including uses, its economic importance, health effects, exposure limits, determination of quantities present in the air, minimization of emissions, labelling, action to take following a spillage and information on formaldehyde in the home.
Published 1993.

—Product Stewardship
Product stewardship is an integral part of Responsible Care and is defined as 'the responsible and ethical management of the health, safety and environmental aspects of a product from its invention, through its processes of production, to its ultimate use and beyond'. This guidance booklet describes how product stewardship can be applied to each aspect of the product lifecycle from research and development through manufacturing, sales and distribution to use and disposal.
Published 1992.

—Protection of the Eyes
This booklet has been produced by the Medical Working Group of CIA to assist companies in their responsibilities to provide safe working conditions in respect of eye protection. It covers eye hazards and the principles of protection from liquids, vapours, gases, particulates and radiation. The guidance includes recommendations on personal protective equipment, contact lenses, training and pre-employment examinations. It also covers legal requirements, treatment of eye injuries and British Standards for eye protection.
First published 1963; Third Edition 1990.

—Responsible Care
Responsible Care is about performance. It is the UK chemical industry's commitment to continuous improvements in all aspects of health, safety and environmental protection. Responsible Care is a voluntary programme of action, fundamental to the industry's present and future performance and the key to maintaining public confidence and acceptability.

In July 1992, all members of CIA committed themselves to upholding the principles of Responsible Care. This brochure describes the guiding principles of Responsible Care, how it is being implemented and how it will affect the chemical industry, its employees, the environment and the public.
Published 1992.

—Safe Handling of Colourants
This publication gives recommendations to users for ensuring the safe handling and use of industrial colourants in the practical industrial context. It deals primarily with potential risks to health although safety considerations are also briefly addressed. It is *not* intended to be a comprehensive guide to the COSHH Regulations.
Published 1989.

—Safe Handling of Potentially Carcinogenic Aromatic Amines and Nitro-Compounds
This focuses on occupational hygiene strategies for controlling potentially carcinogenic aromatic amines and nitro-compounds used in the workplace. Written by a team of occupational health and safety specialists, it is intended for use by non-toxicologists who have some familiarity with toxicology and occupational hygiene practices. It describes a scheme for ranking the carcinogenic potential of aromatic amines and nitro-compounds. The scheme is designed to be used in conjunction with workplace hygiene control strategies which are also described in the document.
Published 1992.

—Safety, Occupational Health and Environmental Protection Auditing
This document includes a revised text of the safety audits guide, published in 1973, in the light of experience gained over the years, and completely new guidance on the auditing of occupational health practice and environmental protection of businesses in the chemicals and related industries, on an integrated basis from both management and technical viewpoints.
Published 1991.

Appendix II

Details of these and many other CIA publications are available from:

CIA Publications,
Kings Buildings,
Smith Square,
London, SW1P 3JJ, UK.
(Tel: 071-834 3399; Fax 071-834 4469)

Index

Page numbers for the main treatment of a subject are given in bold type.

academic institutions, 53–54
AGGIH, 24, 54, 102, 134, 158, 159
acne, 231
ACoPs, 4–6
   carcinogens, 5
   fumigation, 6
   general, 4–5
   Legionnaire's disease, 6
   non-agricultural pesticides, 6
   vinyl chloride, 5–6
action levels, 161
ACTS, 7
acute poisoning, 213–224
   by inhalation, 215
   by ingestion, 215
   by skin contact, 216
   incidents, 221
      planning, 221–222
   recognition and diagnosis, 216
   records, 222–223
   symptoms, 217
   treatment, 217–218
acute toxicity, 98–99
ADs, 235
ADIs, 45, 248, 264
adsorbents, for poisons, 219
AFOM, 63, 64
AHH test, 234
AIHC, 54
AM, 186
American legislation, 34–35

angiosarcoma, 255
animal data, extrapolation, 101
antagonism, 102
antidotes, 219–220
APME, 52
ARAB, 22
Arbo Act, 27
asbestos regulations, 11
asphyxiants, 215
assessments
   of hazards, *see* hazards, assessments
   of risks, *see* risk assessments
auditing, 71–94
   inspections, 83
   management, 84–85
   performance rating, 78
   records, 78–79
   results, 79–82
audit teams, 77
Authorised and Approved List, 11, 120

BAA, 52
base set tests, *see* new chemicals
BATs, 161
Bayer PLC, 105, 113–115
BEIs, 158, 161
BELs, 87
Belgian legislation, 22–26
BGs, 27
BIAC, 46–47
BIBRA, 51

BLS, 35
BLVs, 33
BOHS, 122
BP, 219
breathing apparatus, 147
BRMA, 52, 112
BSI, 54, 135

cancer, *see* carcinogens
carcinogenicity, 100
   guidelines, 8
   tests, 100
carcinogens, 9, **255–259**
   ACoP under COSHH, 5
   IARC classification, 45–46
   test animals, 257–258
   test materials, 258
   test procedures, 258–259
CEC, 28
CEFIC, **49–50**, 176
cell necrosis, 246
CEN, 54, 141, 176, 182, 206, 208
CFCs, 50, 248
CFR, 35
chemical inventories, 75
chemical protective clothing, 9, 15, 87, 92, 122, 137–156, 141–142, **150–154**
   maintenance, 154
   materials, 151–152
   penetration, 150–151
   permeation, 150–152
   purchase and supply, 155
   selection, 152–154
   styles, 152
   training for, 154–155
chemicals control, 248–252
chloracne, 233–234
chromosomes, 253
chronic health effects, 243–267
chronic toxicity, **99–100**, 246–248
CIA, **52**, 115
CIIT, 54
cirrhosis, 246
CISHEC, 52
Classification, Packaging and Labelling, 11–12, 116, 233, 248–249, 261
CMA, 51
CNS, 98
compensation, Belgium, 26
compressed air line equipment, 147
CONCAWE, 51
containment systems, 128
control measures, 118–122
   case examples, 119–122

   engineering, 127–136
   maintenance, 136
COSHH, *see* Control of Substances Hazardous to Health
Control of Substances Hazardous to Health, **3–11**, 83, 90, 96, 105–112, 119–126, 136, 138–140, 159–160, 234, 235, 237, 263
   assessments, *see* risk assessments
   CIA guidance, 8–9
   HSE guidance, 6–8
   Regulations, 4
COREPER, 29
corrosivity, 99
Council of Ministers, EC, 29
CPC, *see* chemical protective clothing
CPL, *see* Classification, Packaging and Labelling
CRAM, 28

dangerous pathogens, 7
dangerous preparations, 12
data distribution models, for hygiene surveys, 185–189
delayed neuropathy, 247
dermatitis, 225–241
   contact, 228–230
DIN, 54, 113–115
DNA, 100, 252–253, 256
doctors, 63, 65
dose–effect relationship, 165, 244, 259
downflow booths, 130
Dublin Foundation, 48–49
Dutch legislation, 27

EC directives, 33–34
ECETOC, 50–51
EC legislation, 28–34
EC framework directives, 33
Economic and Social Committee, 29
eczema, 227–228
EH 40, 7, 120
EH 42, 7, 111, 195
EHC documents, 44, 50
EINECS, 14, 249, 251–252
ELINCS, 15
EMAs, 235
EMAS, 3, 59
ENs, 63, 64
EP, 29
EPA guidelines, 260
epidemiology, 259
ERL, 23
ESC, 29
eukaryotes, 252

EUROMETAUX, 52
European Commission, 28–29
European Community legislation, 28–34
European Court of Justice, 29
European Inventory (EINECS), 14
European Parliament, 29
exhausts, treatment, 135
existing chemicals, 249, **251–252**
exposure
   characteristics, 97–98
   control, 243–267
      auditing, 92–93, 94
   duration, 103
   monitoring, 111
   records, Belgium, 25
   routes of entry, 102, 103, **116–117**, 214
   to mixtures and preparations, 101–102
exposure limits, 7, 8, 75, 87, 94, 103, **157–171**
   application of, 168
   definitions, 160–161
   for mixtures, 9, 168, 183–185
   health-based, 160
   misunderstandings, 169–170
   setting of, 161–167
   working limits, 160–161
extract systems, 131–134
   portable, 134
eyes regulations, 12

FAO, 45
fertility, 262
first aid, 12, 218, **220–222**
French legislation, 28
fume cupboards, 135
fumigation ACoP under COSHH, 6

gametes, 253, 261
gassing incidents, 214
general practitioners, 63, 65
generic assessments, 109–111
genes, 252
German legislation, 27
germ cells, 253
GIFAP, 51
GLP, 47, 256
GM, 186, 194
GMC, 64
GPs, 63, 65
GSD, 187, 194–197

hazards
   assessments, **95–104**, 139–140, 161, 245
   definition, 106, 139, 245
   evaluation, 101–104

information, 8
symbols, 13, 116
HAZOPs, 92, 111
Health and Safety Commission, 3
Health and Safety Executive, 3
health auditing, 71–94
health-based limits, 160
health services, 57–70
health standards, 74–77
health surveillance, 6, 9, **234–236**
HEDSET, 48
HMFI, 59
HPV, 251
HSE guidance
   EH 40, 7
   EH 42, 7
   under COSHH, 6
HSGs, 44
HSW Act, 3, 61
hypertrophy, 246

IARC, 45–46
ICCA, 20, **51**
ICOH, 48
ICSCs, 20, **44**
IgE, 239
IIAC, 14, 52
ILO, 40–42
   conventions and recommendations, 58
      table, 41
ILVs, 33
IM injections, *see* injections, for poisoning
IMO, 99
industry organizations, international, 49–52
information, instruction and training, 9
injections, for poisoning, 220, 223–224
injuries, reporting regulations, 13
international organizations, 40–52
   industry supported, 49–52
international regulations, 19–38
IOSH, 122, 123
IPCS, 43–45
irritancy, 99, 215, 233
ISO, 54, 141
Italian legislation, 28
IV injections, *see* injections, for poisoning

JECFA, 45
JMPR, 45

laboratories, 9
laminar flow booths, 130–132
latent period, 256
LC 50 test, 99

LD 50 test, **98–99**, 217
lead, 11
Legionnaires' disease, ACoP under COSHH, 6
legislation
  Belgium 22–26
  European Community, 28–34
  France, 28
  Italy, 28
  Germany, 27
  The Netherlands, 27
  UK, 1–18
  USA, 34–38
LEV, 129–135
literature sources, 115–116
local exhaust ventilation, 129–135
long-term toxicity, 99–100

MACs, 159
MAK Commission, 27
MAK values, 27, 159, 161
maximum tolerated dose, 258
MDHS, 182
mechanical ventilation, 129
medical examinations, Belgium, 25–26
medical records, 69
MEDICHEM, 48
meiosis, 253
MELs, 5, 7, 120, **125**, 159–160, 161, 179, 185, **203**
metabolites, 246
MFOM, 63, 64
MDHS, 182
mitosis, 253
mixed exposures, 101–102
monitoring, 7, **236–237**
MTD, 258–259
mutagenicity, 100–101, **252–254**

NADOR, 13
NAMAS, 182
narcotic agents, 215
national institutions, 53
natural ventilation, 128–129
neuropathy, delayed, 247
neurotic esterases, 247
new chemicals
  base set tests, 250
  notification regulations, 14–15, **249–251**
NIOSH, 35, 196
no-effect levels, 104, **166–167**, 244, 264
NOEL, *see* no-effect levels
nominal protection factor, 121
non-agricultural pesticides, ACoP under COSHH, 6
non-governmental organizations, 39–55
NONS, *see* new chemicals
notification regulations, *see* new chemicals
NPF, 121
NTP, 258

occupational diseases, Belgium, 25
occupational health
  auditing, 71–94
  definition, 58
  nurses, 62–63, **64**
  team, 66
  third party, 65
occupational health services, 57–70
  levels, 63, 66
  objectives, 59–60
  provision of, 62–63
  recruitment, 66–68
  setting up, 63–68
occupational hygiene surveys, 175–183
  basic surveys, 177–179
  detailed surveys, 179–180
  quality, 182
  routine surveys, 180–182
  sampling and analytical methods, 183
occupational physicians, 64
OECD, 20, **46–48**, 245, 248
  test guidelines, 47, 250, 259
OELs, *see* exposure limits
OESs, 7, **126**, 159–160, 161, 185, **203**
OHNC, 63, 64
OHND, 63, 64
OHSs, 57–70
OJ, 31
organizations
  international, 40–52
  non-governmental, 39–55
  professional, 54
organophosphate poisoning, 223–224
OSHA, 21, **34–38**
  standards, 35–38
  table, 36–37
pathogens, dangerous, 7
PCIAOH, 48
PELs, 37
personal protective equipment, *see* chemical protective clothing
pesticides, non-agricultural, 6
pH, 232
poisoning, *see* acute poisoning
poisons centres, 224
PPE, *see* chemical protective clothing
preparations, dangerous, 12

prescribed diseases, 14
probability plots, **187–190**, 194, 205
professional organizations, 54
prokaryotic cells, 252
protection factors, 148
  nominal, 121
protective clothing, see chemical protective clothing
proteins, 252

QMV, 29

R 45, 5
R 49, 5
RAST test, 239
RCN, 64
RCP, 64
record keeping, 9
  and computerization, 123–125
regulations, UK, 1–18
regulatory systems, 20–22
reporting of injuries regulations, 13
reproductive toxicity, 101, 261–262
respirators, 143–145
  filters, 144–145
respiratory protective and equipment, 7, **142–150**
  maintenance, 150
  selection, 147–148, 149
  standards, 141
Responsible Care, 9, 10
  health indicator, 94
RGN, 63, 64
RGPT, 22
RICE, 182
RIDDOR, 13–14, 59, 90, 222, 240
'rip and tip' machines, 134
risks, definition, 106, 139, 245
risk assessments, 8, 60–61, 75, **105–126**, 140, 164–166, 234, 245, **263–266**
  area-based, 107
  control-based, 107
  generic, 109–111
  mathematical modelling, 264–266
  process-based, 107
  results, 123
  safety factor approach, 263–264
  substance-based, 106–107
RPE, see respiratory protective equipment
RSC, 107, 123

safety data sheets, 111–112, 113–115
safety factors, 263–264
sampling strategies, 173–211

biological considerations, 189–191
  measurement strategies, 197–202
  results, interpretation, 203–209
sample populations, 191–197
SCBA, 147
SEA, 28, **32–33**
sensitization, skin, 100, 233
Seventh Amendment, 249–261
SF, see safety factors
short-term screening tests, 253–254
short-term toxicity, 99
SIDS, 47
Single European Act, 28, **32–33**
Sixth Amendment, 249
skin diseases, 227–241
  management, 240–241
  notifiable, 238, 240
skin sensitization, 100
skin
  biopsy, 239
  blood tests, 239
  cancer, 231, 233–234
  patch testing, 239
  prick testing, 238–239
  role and function, 226–227
somatic cells, 253
SOPs, 93
standards, for PPE, 140–142
standards institutions, 54
STELs, 197–198
sub-acute toxicity, 99
sub-chronic toxicity, 99
substances, identification, 97
synergism, 102

technical progress committees, EC, 29
teratogenicity, 101, **259–262**
TLVs, 24, 54, 158, 159
toxicity data, review, 98–101
toxicokinetics, 262
toxicology, 96, 244, **252–262**
TQM, 26, 54
trade associations, 112, 115
TRK values, 27, 159, 161
TWA, 189, 191

UK regulations, 1–18
ulceration, 231
UNEP, 43
UNICE, 48, 50
universities, 53
UN organizations, 40–46
urticaria, 230
US legislation, 34–38

ventilation systems, 127–136
  local exhaust, 129–135
  natural, 128–129
vinyl chloride, ACoP under COSHH, 5
vitiligo, 230, 233–234
vomiting, induction of, 218

WASP, 182
WATCH, 7
WHO, 42–43, 176, 262
WTR, 52

ELLIS HORWOOD SERIES IN APPLIED SCIENCE AND INDUSTRIAL TECHNOLOGY

Series Editor: Dr D. H. SHARP, OBE, former General Secretary, Society of Chemical Industry; former General Secretary, Institution of Chemical Engineers; and former Technical Director, Confederation of British Industry

series continued from front of book

QUALITY MANAGEMENT IN THE SERVICE INDUSTRY
L. STEBBING, Quality Management Consultant
REFRACTORIES TECHNOLOGY: Manufacture and Industrial Application
C. STOREY, Refractories Consultant
INDUSTRIAL PAINT FINISHING TECHNIQUES AND PROCESSES
G.F. TANK, Educational Services, Graco Robotics Inc., Michigan, USA
MODERN BATTERY TECHNOLOGY
Editor: C.D.S. TUCK, Alcan International Ltd, Oxon
FIRE AND EXPLOSION PROTECTION: A Systems Approach
D. TUHTAR, Institute of Fire and Explosion Protection, Yugoslavia
COSMETICS AND TOILETRIES: Development, Production and Use
Editor: WILFRIED UMBACH, Director, Chemical Research, Henkel KGaA, Dusseldorf, Germany
PERFUMERY TECHNOLOGY 2nd Edition
F.V. WELLS, Consultant Perfumer and former Editor of *Soap, Perfumery and Cosmetics*, and
M. BILLOT, former Chief Perfumer to Houbigant-Cheramy, Paris, Président d'Honneur de la Société Technique des Parfumeurs de la France
THE MANUFACTURE OF SOAPS, OTHER DETERGENTS AND GLYCERINE
E. WOOLLATT, Consultant, formerly Unilever plc
HOW TO TURN ROUND A MANUFACTURING COMPANY
BRIAN HALFORD WALLEY, Former Finance and Site Director, Ferodo Ltd